AF848029

Information Systems Engineering and Management

Volume 37

Series Editor

Álvaro Rocha, ISEG, University of Lisbon, Lisbon, Portugal

Editorial Board

Abdelkader Hameurlain, Université Toulouse III Paul Sabatier, Toulouse, France

Ali Idri, ENSIAS, Mohammed V University, Rabat, Morocco

Ashok Vaseashta, International Clean Water Institute, Manassas, VA, USA

Ashwani Kumar Dubey, Amity University, Noida, India

Carlos Montenegro, Francisco José de Caldas District University, Bogota, Colombia

Claude Laporte, University of Quebec, Québec, QC, Canada

Fernando Moreira, Portucalense University, Berlin, Germany

Francisco Peñalvo, University of Salamanca, Salamanca, Spain

Gintautas Dzemyda, Vilnius University, Vilnius, Lithuania

Jezreel Mejia-Miranda, CIMAT - Center for Mathematical Research, Zacatecas, Mexico

Jon Hall, The Open University, Milton Keynes, UK

Mário Piattini, University of Castilla-La Mancha, Albacete, Spain

Maristela Holanda, University of Brasilia, Brasilia, Brazil

Mincong Tang, Beijing Jaiotong University, Beijing, China

Mirjana Ivanovíc, Department of Mathematics and Informatics, University of Novi Sad, Novi Sad, Serbia

Mirna Muñoz, CIMAT Center for Mathematical Research, Progreso, Mexico

Rajeev Kanth, University of Turku, Turku, Finland

Sajid Anwar, Institute of Management Sciences, Peshawar, Pakistan

Tutut Herawan, Faculty of Computer Science and Information Technology, University of Malaya, Kuala Lumpur, Malaysia

Valentina Colla, TeCIP Institute, Scuola Superiore Sant'Anna, Pisa, Italy

Vladan Devedzic, University of Belgrade, Belgrade, Serbia

The book series "Information Systems Engineering and Management" (ISEM) publishes innovative and original works in the various areas of planning, development, implementation, and management of information systems and technologies by enterprises, citizens, and society for the improvement of the socio-economic environment.

The series is multidisciplinary, focusing on technological, organizational, and social domains of information systems engineering and management. Manuscripts published in this book series focus on relevant problems and research in the planning, analysis, design, implementation, exploration, and management of all types of information systems and technologies. The series contains monographs, lecture notes, edited volumes, pedagogical and technical books as well as proceedings volumes.

Some topics/keywords to be considered in the ISEM book series are, but not limited to: Information Systems Planning; Information Systems Development; Exploration of Information Systems; Management of Information Systems; Blockchain Technology; Cloud Computing; Artificial Intelligence (AI) and Machine Learning; Big Data Analytics; Multimedia Systems; Computer Networks, Mobility and Pervasive Systems; IT Security, Ethics and Privacy; Cybersecurity; Digital Platforms and Services; Requirements Engineering; Software Engineering; Process and Knowledge Engineering; Security and Privacy Engineering, Autonomous Robotics; Human-Computer Interaction; Marketing and Information; Tourism and Information; Finance and Value; Decisions and Risk; Innovation and Projects; Strategy and People.

Indexed by Google Scholar. All books published in the series are submitted for consideration in the Web of Science.

For book or proceedings proposals please contact Alvaro Rocha (amrrocha@gmail.com).

Rathika Palanisamy · Azah Anir Norman ·
Miss Laiha Mat Kiah · Tutut Herawan

Bring Your Own Device Security Policy Compliance Framework

Springer

Rathika Palanisamy
Information Technology Division (ITD)
Ministry of Finance
Putrajaya, Malaysia

Miss Laiha Mat Kiah
Department of Computer System
and Technology
Faculty of Computer Science
and Information Technology
University of Malaya
Kuala Lumpur, Malaysia

Azah Anir Norman
Department of Information Systems
Faculty of Computer Science
and Information Technology
University of Malaya
Kuala Lumpur, Malaysia

Tutut Herawan
Department of Information Systems
Faculty of Computer Science
and Information Technology
University of Malaya
Kuala Lumpur, Malaysia

ISSN 3004-958X ISSN 3004-9598 (electronic)
Information Systems Engineering and Management
ISBN 978-3-031-86800-9 ISBN 978-3-031-86801-6 (eBook)
https://doi.org/10.1007/978-3-031-86801-6

© The Editor(s) (if applicable) and The Author(s), under exclusive license to Springer Nature Switzerland AG 2025

This work is subject to copyright. All rights are solely and exclusively licensed by the Publisher, whether the whole or part of the material is concerned, specifically the rights of translation, reprinting, reuse of illustrations, recitation, broadcasting, reproduction on microfilms or in any other physical way, and transmission or information storage and retrieval, electronic adaptation, computer software, or by similar or dissimilar methodology now known or hereafter developed.
The use of general descriptive names, registered names, trademarks, service marks, etc. in this publication does not imply, even in the absence of a specific statement, that such names are exempt from the relevant protective laws and regulations and therefore free for general use.
The publisher, the authors and the editors are safe to assume that the advice and information in this book are believed to be true and accurate at the date of publication. Neither the publisher nor the authors or the editors give a warranty, expressed or implied, with respect to the material contained herein or for any errors or omissions that may have been made. The publisher remains neutral with regard to jurisdictional claims in published maps and institutional affiliations.

This Springer imprint is published by the registered company Springer Nature Switzerland AG
The registered company address is: Gewerbestrasse 11, 6330 Cham, Switzerland

If disposing of this product, please recycle the paper.

I owe great amount of gratitude to my husband Cdr. Dilagar Muthumani and my children, Tarun Aditya, Darshana Shasvati and Trilokesh Ananda who are my biggest blessing and my pillar of support. Without their tremendous understanding, support and encouragement, this study would have been impossible. Lastly, to my precious little boy, Jayaviir Ananda whom I shared this struggle with, raising him while doing my Ph.D. has taught me determination, the word "good enough", to embrace imperfection and celebrate small victories.

Rathika Palanisamy

All praise to Allah for His endless guidance. To my voices of comfort, Sadiah Mustajab and Norman Ibrahim, thank you for your love and prayers. To my confidant, Mohd Anuar Mustafa, and my qorrata a'yun—Ain Zahirah, Afiah Zaheen, Auni Zakiyah, Ahmad Zuhayr, and Anis Zahrah—your support and love are my greatest strength.

Azah Anir Norman

For everything—Allah. For my life—Siti Rahmah and Mat Kiah. For my strength—Nor Asrul. For my motivation—Danial Irfan and Zarin Sophia.

Miss Laiha Mat Kiah

This book is dedicated to my beloved family: Nau Herawan, Cherry Herawan, and Sya Herawan.
The help of my precious wife Eka Novita Sari is highly appreciated in the completion of this book.

Tutut Herawan

Preface

Proliferation of Bring Your Own Device (BYOD) has instigated a widespread change, fast outpacing the security strategies deployed by organizations. The influx of these devices has created information security challenges within organizations, further exacerbated with employees' inconsistent adherence with BYOD security policy. To prevent information security breaches, compliance with BYOD security policy and procedures is vital. The *Bring Your Own Device Security Policy Compliance Framework* book aims to determine the factors that employees' BYOD security policy compliance by using mixed methods approach. Security policy compliance factors, BYOD practices and security risks were identified following a systematic review approach. Building on Organizational Control Theory, Security Culture and Social Cognitive Theory, a research framework positing a set of plausible factors determining BYOD security policy compliance was developed. Next, with a purposive sample of eight information security experts from selected public sector organizations, interviews and BYOD risk assessments analysis were performed to furnish in-depth insights into BYOD risks, its impact on organizations and recommend control measures to overcome them. This led to the suggestion of four control measures to mitigate critical BYOD security risks such as Security Training and Awareness (SETA), policy, top management commitment and technical countermeasures. The control measures were mapped into the research framework to be tested in the following quantitative phase. The proposed research framework was tested using survey results from 346 employees of three Critical National Information Infrastructure (CNII) agencies. Using Partial Least Squares—Structural Equation Modelling (PLS-SEM), the framework's validity and reliability were evaluated, and hypotheses were tested. Findings show that perceived mandatoriness, self-efficacy and psychological ownership are influential in predicting employees' BYOD security policy compliance. Specification of security policy is associated with perceived mandatoriness, while BYOD IT support and SETA are significant towards self-efficacy. Unexpectedly, security culture has been found to have no significant relationship to BYOD security policy compliance. Theoretical, practical and methodological contributions were discussed and suggestions for future research were recommended. The analysis led to a number of insightful findings that contribute to the literature and the management, which are predominantly centered on traditional computing. In view of the ever-increasing BYOD threats to the security of government information, it is imperative that IT managers establish and implement effective policies to protect vital information assets. Consequently, the findings of this study may benefit policy-makers, particularly in the public sector, in their efforts to increase BYOD security policy compliance among employees.

Kuala Lumpur, Malaysia

Rathika Palanisamy
Azah Anir Norman
Miss Laiha Mat Kiah
Tutut Herawan

Acknowledgements The authors would like to extend their heartfelt gratitude and appreciation to the Ministry of Higher Education (MOHE), Malaysia, for funding this project with the Fundamental Research Grant Scheme (FRGS) (FP056-2019A). Ref. no.: FRGS/1/2019/ICT04/UM/02/1.

Contents

About the Authors

Rathika Palanisamy is attached to the Information Technology Division, Ministry of Finance. She completed her doctoral degree at the Department of Computer Systems and Technology, Faculty of Computer Science and Information Technology, University of Malaya, Malaysia, in 2023. Her research interests include information security risk management and security policy compliance in organizations.

Azah Anir Norman is an associate professor and currently the Deputy Dean of Value Creation and Enterprise, Faculty of Computer Science and Information Technology, University of Malaya (UM), Malaysia. She earned her undergraduate degree at Universiti Kebangsaan Malaysia (UKM) and her master's degree in electronic commerce security from Royal Holloway University of London in the UK in 2004. She completed her Ph.D. from the University of Malaya (UM) in 2014. She specializes in information security management systems (ISMS), secure applications for ICT, privacy and human elements in security, information security governance, security on social platforms, and e-commerce security. She is also very interested in topics pertaining to Islamic ICT (such as Halal and Quran Authentication), Design Thinking and Teaching and Learning Innovations. Azah Norman published numerous academic papers in reputable ISI and SCOPUS publications in the fields of information security governance, information security management, information security systems, information security and trust, information security and privacy, information security education awareness, information security and assurance, and information security policy and governance. Before entering the academic world, she worked as a consultant at MSC Trustgate.com, a subsidiary of MDEC and a partner of VeriSign Inc. in the USA. In Trustgate, she provided Internet Security implementation consultation to numerous top 500 companies. Prior to becoming the consultant at Trustgate, she experienced working at VeriSign Inc. in Silicon Valley, San Jose, California, in 2001. As a specialist in information security management systems, she is also part of an expert in the working group WG/G/5-1 Information Security Management System, Department of Standards, Malaysia, and the International Organisation for Standardisation (ISO). She belongs to the Association of Information Systems (AIS) and the MyAIS (AIS Malaysia Chapter), an organization that promotes excellence and knowledge progress in the field of information systems research and practice. She is also a secretary at the Cybersecurity Academia Malaysia Association (CSAM), a national association that promotes cybersecurity teaching, awareness and research in Malaysia. She received a prestigious award from the Royal Academy of Engineering of the United Kingdom (RAENG) as the Leader of Innovation in 2018.

Miss Laiha Mat Kiah received her Ph.D. degree in Information Security from Royal Holloway, University of London, United Kingdom, in 2007, and since then she has been an academic and an active researcher at the Faculty of Computer Science and Information Technology, Universiti Malaya (UM), Kuala Lumpur, Malaysia. Her fundamental discipline is Computer Science, and her area of expertise is Cyber Security (and its related topics). She was promoted to the full Professorship in 2015 and is an active member of Malaysia Board of Technologists (Ts.), Malaysian Society for Cryptology Research (MSCR), IEEE as Senior Member, and EC Council Member. Her main research interest will always be in the security aspect of computing and technology fields with variation of applications in multi and/or trans-disciplinary projects. This is evidenced by her publications and research projects in which she is/was the principal investigator (PI) as well as co-PIs. As a professional technologist (Ts.), keeping up with the current trend and demand of ever evolving computing technology field is crucial to ensure the quality and the impact of her research work. Current research interests include cyber security, blockchain technology, IoT and health information exchange.

Tutut Herawan is an associate professor at the Department of Information Systems, Faculty of Computer Science and Information Technology, Universiti Malaya, Malaysia. He was named in the Top 2% World Scientists Ranking by Stanford University and Elsevier BV, since 2019-present. He received a Ph.D. degree in information technology from Universiti Tun Hussein Onn Malaysia in 2010. He has more than 18 years experiences as academic staff and has supervised many Master and Ph.D. students. He is the associate editor of many international journals of Scopus and Clarivate Web of Science indexes. He has edited many books and published more than 330 articles in various book chapters, international journals, and conference proceedings. He is an active reviewer for many reputable international journals. He delivered more than 50 keynote addresses, invited workshops and seminars and has been actively served as a chair, co-chair, program committee member and co-organizer for numerous international conferences/workshops. His research area includes data science and big data, data engineering and information systems, soft computing and data mining, and information technology for tourism.

Symbols and Abbreviations

AVE	Average Variance Extracted
BYOD	Bring Your Own Device
CERT	Computer Emergency Response Team
CFA	Confirmatory Factor Analysis
CMB	Common Method Bias
CNII	Critical National Information Infrastructure
DV	Dependent Variables
GDP	Gross Domestic Product
GDT	General Deterrence Theory
HBM	Health Belief Model
HOC	Higher-Order Constructs
HTMT	Heterotrait-Monotrait Ratio
ICT	Information Communication and Technology
IS	Information System
ISO/IEC	International Organization for Standardization (ISO)/International Electrotechnical Commission (IEC)
ISP	Information Security Policy
IT	Information Technology
IV	Independent Variables
LOC	Lower-Order Constructs
MAMPU	Malaysian Administrative Modernisation and Management Planning Unit
NACSA	National Cyber Security Agency
NCSP	National Cyber Security Policy
NITC	National Information Technology Council
OCT	Organizational Control Theory
PC	Psychological Contract
PCT	Psychological Contract Theory
PLS-SEM	Partial Least Squares-Sequential Equation Modelling
PMT	Protection Motivation Theory
PPT	People, Process and Technology

RT	Reactance Theory
SCT	Social Cognitive Theory
SETA	Security Education, Training and Awareness
SLR	Systematic Literature Review
SME	Subject Matter Expert
TPB	Theory of Planned Behavior
TTAT	Technology Threat Avoidance Theory

List of Figures

List of Tables

Chapter 1
Introduction

1.1 Background

The Bring Your Own Device (BYOD) practice in workplaces has fundamentally transformed the work style, culture, and technology landscape of organizations. The market for BYOD is projected to reach $430.45 billion by 2025, up from $186 billion in 2019 [72]. In certain research, BYOD is also referred to as IT consumerization [106], which is the result of the changing role of technology and the employee/consumer-origin technologies that are brought into the workplace [54]. The widespread acceptance of cutting-edge mobile phone technology, as well as constant advances in the area has resulted in an exponential expansion in the smartphone market [74, 117]. BYOD has altered the way work is performed in organizations [82] as employees use their own devices to access online services such as cloud storages and social media on the same devices.

There are numerous benefits of BYOD to individuals and enterprises, ranging from cost savings to increased productivity, adaptability, and efficiency of personnel [118]. BYOD allows employees to access corporate data, controlled networks, applications, and records, as well as the Internet, using their own mobile devices. In this way, employees are privileged and empowered to choose the technology that best match their specific needs [77]. BYOD enhances mobility, flexibility, productivity, and employee satisfaction by extending the opportunity to work whenever and wherever [49, 51, 98]. According to studies, the use of personal devices to complete work faster and better customer service, results in increased revenue for organizations.

The Malaysian public sector has embraced BYOD as part of the working culture over the last decade [73]. The extensive use of BYOD among government professionals is primarily due to the acceleration of digital services and the rapid advancement of disruptive technologies that can significantly transform the economy landscape of the country [78]. The explosive growth of remote working or telecommuting due to the Covid-19 pandemic [34, 87, 123] may not be feasible without

© The Author(s), under exclusive license to Springer Nature Switzerland AG 2025

R. Palanisamy et al., *Bring Your Own Device Security Policy Compliance Framework*, Information Systems Engineering and Management 37,
https://doi.org/10.1007/978-3-031-86801-6_1

BYOD. For this reason, BYOD has further expanded since the advent of the Covid-19 pandemic. In order to accommodate the growing number of remote employees, organizations have already adapted for the future by allowing their employees to access the corporate network and data via personal devices such as smartphones, laptops, and tablets [46]. Additionally, the pandemic has also pushed the public sector for the digital government transformation, where all citizen centric services are expected to continue seamlessly and provide effective services in a timely manner to the public [42, 103].

The BYOD trend will increase as more digital natives enter the workforce and expect to use their mobile devices at work [9, 119]. The increasing demand for mobile devices in the Asia–Pacific area is a result of the region's growing number of tech-savvy individuals. By 2027, smartphone users are projected to utilize 7.7 billion subscriptions, up from an estimated 6.23 billion in 2021 [107]. Approximately 44% of Millennials or Generation Y, also known as digital natives, are entering the workforce with differing perspectives on technology and how work should be performed using cutting-edge technologies [61, 65]. This generation, which represents the digital workforce, accelerates the consumerization of the workplace, which influences how organizations view technology's role in the workplace. They demand their employers to permit them to carry personal electronic devices such as smartphones, tablets, and laptops to work resulting in the organizational perimeter being expanded.

The proliferation of these devices has presented organizations with new, unpredictable, and unanticipated information security challenges [8, 25, 104]. The decision to adopt digital technology not only opened the door to new opportunities, but also to threats, risks, and vulnerabilities [95] which is quickly superseding the security measures outlined by the organizations. In the public sector, BYOD practice is one of the major causes for data leakages as devices are at an increased risk of breaches. Performing work related task using personal devices with official information and secrets stored in mobile devices results in data privacy and security implications.

The exponential growth of BYOD highlights and establishes the necessity for increased focus on information security policy (ISP) compliance by enterprises. This is especially important in organizations, such as the Critical National Information Infrastructure (CNII) in the public sector, that own sensitive information such as medical data, energy, defense, payment, and personal information, as well as intellectual assets [92]. As these organizations are responsible for providing important public services, disruptions to CNII can have a substantial impact on the nation. A security breach in one CNII sector can have repercussions for other key sectors [33].

According to reports, the security policy compliance behavior of employees continues to be a challenge as it is one of the primary sources of organizations' exposure to risk [14, 27, 88, 91, 94, 111, 122]. The majority of threats arise from employees' lack of security compliance [24, 75]. Employees who have access to business infrastructure and data pose one of the greatest cybersecurity risks to a corporation [40]. Access to business infrastructure and data makes employees, one of the greatest cybersecurity threats to an organization. As a result of combining their work and personal lives on the same device, users may unknowingly create information security risks by exposing sensitive data or by introducing additional vulnerabilities that can be exploited. Online businesses collect vast amounts of data from clients' personal devices [39]. Trend Micro surveyed 13,200 remote workers from 27 countries and discovered that 56% admitted to installing and using a non-work application on a corporate device against company policy, 66% uploaded corporate data to a their personal applications, and 39% of respondents "often" or "always" accessed corporate data from a personal device [112].

A compromised BYOD can have far-reaching impact [43] as the failure to address these risks may result in financial and legal implications [97]. Microsoft in collaboration with Frost and Sullivan found that cybersecurity threats may cost organizations USD 12.2 billion in economic losses and that is more than 4% of Malaysia's Gross Domestic Product (GDP) [48]. Information is an organization's most valuable asset, and any mishaps involving it can result in loss of reputation, loss of profits, or loss of intellectual property [99, 100]. In the public sector, it can jeopardize people's trust, national sovereignty, national security, and service delivery [38, 57].

It is typical for employees to use their mobile devices for both work and personal purposes while remaining oblivious of the security policy in place. As humans are considered to be the "weakest link" in any cybersecurity program, human error accounts for most security breaches [17, 18, 55, 94]. Recent study found that the majority of security incidents are caused by people's failure to comply with security policies [37]. Given the extent of the insider threat, ensuring employee ISP compliance is essential to achieving and maintaining information security [6, 7, 76]. As a result, it is critical to control the human element in order to manage information security in businesses [19, 70]. BYOD practices transform the threat landscape of organizations, necessitating new security strategies. Complying to BYOD security policies and procedures is effective to mitigate information security breaches [79, 100, 101, 105, 108]. While it is necessary to design a comprehensive and robust BYOD security strategy [41], it is still inadequate to ensure that employees adhere to it. The success or failure of an organization depends on the security tasks its personnel accomplish or fail to accomplish. Therefore, it is critical to recognize the factors that motivate employees to comply with BYOD security policies [36].

Previous research in this area has relied heavily on Protection Motivation Theory (PMT), relatively neglecting the other dimensions. This clearly shows that it is still an emerging area [86] as research calls for more theories to be applied in this area [13, 64]. There is a need to integrate other theories to improve ISP compliance [58, 67] and human aspects of the security requirements are still underrepresented [28, 85]. In addition, previous research has underlined that technical security approaches alone are insufficient to protect people and organizations [44, 121], therefore individual behavior must be accounted for in security studies [93, 94]. In order to support data-driven applications, information security management must be prioritized [32]. The role of BYOD IT support and psychological ownership in influencing BYOD ISP compliance remains underexplored.

Second, despite the necessity to comprehend BYOD security concerns and elements that lead to security policy compliance behavior, there is a paucity of research on how specific factors were chosen for study as most of the factors were just adopted from extant literature. Understanding these factors could reveal and help build interventions to develop security policy compliance behavior resulting in the reduction of BYOD security risks. Thirdly, the majority of studies collect samples that are not representative of the employee population, especially in the public sector.

There have been a few studies on user information security policy compliance, and the majority of the existing literature has focused on employee information security behavior in a traditional context where the organization owns the assets, leaving a void for research that focuses on specific technologies like BYOD and security policies [47]. Although employees have been identified as one of the greatest threats to an organization's information security, little effort has been made to improve their security policy compliance behavior. The research on employee ISP compliance is still in its infancy [21], since the bulk of the effort to improve information security is centered on technology solutions [83]. Hence, predicting employees' BYOD security policy compliance is an important aspect of security policy compliance. The overall direction of this study is defined by the identification of BYOD security policy compliance factors.

1.2 Definitions and Terms

In this study, these are the definitions employed:

a. **Bring Your Own Device**

A practice in which employees bring their own computing devices to work, such as laptops, smartphones, and/or tablets, and connect them to the organization's network [43, 52]. It is a way of enabling workers to work remotely by granting them access to the organization's network and data from home or while travelling [23].

b. **BYOD Information Security Policy**

Policies define the roles, responsibilities, and standard operating procedures of personnel in relation to the proper use and protection of organizational information system resources [17, 109]. The BYOD security policy would govern how employees access critical organizational data and software on their own devices, including which devices may connect to the corporate network, and it is vital to the prevention of security incidents.

c. **BYOD Security Risks**

Risk is the possibility of losing valuable goods due to noncompliance with BYOD security policy, such as loss of reputation and financial loss. Devices provide a distinct security risk due to their unique characteristics such as mobility and portability [116]. These characteristics make them more vulnerable to data breaches than traditional computing systems. The increasing prevalence of BYOD practices raises the risk, and, if left unmanaged, may provide opportunities for hackers to target devices for their data.

d. **Control measures**

It is the planning and development of ways and solutions to eliminate, minimize, or manage the impact of threats or risks. The mitigation techniques may lessen the chance of the risk, the severity of the consequence, or the exposure to the risk.

e. **Critical National Information Infrastructure**

CNII is the key infrastructure that supports the economic, political, strategic, political, and socioeconomic operations of the nation [30]. The CNII are made up of ten sectors namely: National Defense and Security; Banking and Finance; Information and Communications; Energy; Transportation; Water; Health Services; Government; Emergency Services; and Food and Agriculture [33]. The destruction or disruption of these systems and communication networks would have a severe impact on the nation's economic strength, image, defense and security, government's operational capacity, and public health and safety [124].

f. **Cyber attack**

An actual assault by an intentional threat source that aims to either alter a system, its data or its operation (active attack) or learn or use information from a system, but does not attempt to affect the system, its resources, its data, or its operation (passive attack). Control system assaults, phishing attacks (including personalized spear phishing attacks), malware, ransomware, Trojan horses, unauthorized access, viruses, and worms are only a few of the many sorts of cyber threats.

g. **Data breaches**

Data breaches are any activities that result in the disclosure of secure, confidential data to unreliable technologies and environments in order to expose organizations' vulnerabilities [50].

h. **Data leakage**

Data leakage is the unauthorized transfer or compromise of data from within an organization or internal environment to external sources. It is mostly caused by insiders who share confidential data with unauthorized recipients, or leave a gap that enables that information to be accessed by unauthorized people.

i. **End user**

The individual for whom products or policies are created. In the context of organizational cybersecurity, the end user is any employee who interacts with the organization's information communication technology, and they are unlikely to have the technical competence required to properly understand the technology they are using.

j. **Information Security Policy (ISP) compliance**

An information security compliance behavior is characterized by actions that do not violate an organization's ISP. It is the degree to which employees adhere to these organizational policies in order to use IS appropriately on the job. Compliance with security policies indicates that personnel have met a predefined set of security standards specified at the time. People are the most vulnerable component of the cybersecurity infrastructure; regardless of how advanced an organization's technology is, security will fail if employees do not adhere to security regulations.

k. **Information Systems (IS) Security**

IS security consisting of a wide array of operations designed to control and manage potential data and information threats.

l. **Insider threat**

Possibility that a person who has or has had authorized access to an organization's assets would use that access to behave in a way that could have a negative impact on the organization, whether maliciously or accidentally [26]. There are two types of insider threats; malicious and unintentional insider threat. A malicious insider threat is a current or former employee, contractor, or business partner who has or had authorized access to an organization's network, system, or data and intentionally exceeded or misused that access in a way that negatively affected the confidentiality, integrity, or availability of the organization's information or information systems. An unintentional insider threat is a current or former employee, contractor, or business partner who has or had authorized access to an organization's network, system, or data and who, through action or inaction without malice, causes harm or significantly increases the probability of future serious harm to the confidentiality, integrity, or availability of the organization's information or information systems [20, 122].

m. **Malaysian Public Sector**

Federal, state, and municipal governments comprise the Malaysian Public Sector's three tiers. In carrying out its responsibilities, the federal government establishes several ministries, departments, and agencies (statutory bodies). State government are the second layer of the government hierarchy, whereas the local government are the lowest tier [12, 110].

n. **Organizational Information Security**

The use of fundamental risk principles to manage all security risks, whether they are connected to information, cyber, physical security, asset management, or business continuity, in a comprehensive, holistic, all-encompassing manner.

o. **Phishing attack**

A sort of social engineering attack in which victims are driven to a fake website via a link, which is typically found in an email or other communication sent to them or on websites they may visit. When the link is activated (website access or file download), it may include malware or trigger additional messages asking for personal information from the user [31].

p. **Privacy**

Individuals' capacity to comprehend and exercise control over how information about them may be utilized by other parties.

q. **Risk**

The probability that an incident, event, or occurrence will end in an undesirable or unfavorable outcome, as indicated by the likelihood that a specific vulnerability will lead to the corresponding consequences.

r. **Risk assessment**

The product or process that collects information and assigns values to risks in order to determine priorities, develop, or compare courses of action, and inform decision making.

1.3 Research Problem

The proliferation of mobile devices has altered workplace technology and inevitably allow employees to hop between corporate tools and personal applications on their devices. These practices introduce many information security risk to the organizations [2]. Over 93% of organizations report that they face issues with employees using unauthorized personal devices and storage for corporate data [84]. Research also found that employees induce 60% of the security breaches in the organization [90]. Risky employee behaviors include, among others, the use of unapproved apps, the misuse of corporate computers, unauthorized physical and network access, remote employee security, and the misuse of passwords [1, 3, 4]. About 30% of the organization lacks visibility and control over personal devices, giving potential for cybercriminals to breach organizational data [71].

Malaysia is the eleventh most breached country in the first half of 2022, with the public sector agencies being continuously under attack by cyber criminals. In November, 2022, data breach had occurred at the Election Commission, resulting in a database containing the details of over 800,000 Malaysians being sold online for a mere US$2000 [89]. A few months before this incident, millions of datasets belonging to the National Registration Department (NRD) were up for sale for just US$10,000 (RM44,095). Most recently, there were attacks by hackers to steal military secrets, personal information from the infected devices including passwords, browser history, and data from social apps like Viber and Telegram [16].

The pandemic has accelerated and normalized remote working for millions of people around the world [35]. BYOD has made the transition easier as employees were forced to work from home overnight. At the same time, this transition has also led to a high wave of cybercrimes targeting remote workers [68, 112, 123]. Risk is multiplied when remote employees use less protected personal devices to access organizational data [53] and rely on home networks to complete tasks. According to research, there has been a 250% surge in cybercrime aimed at work-at-home employees, particularly nation-state threats from Russia, Iran, North Korea, and China [120]. The lack of understanding to detect and battle possible BYOD risks hinders the organization's capacity to improve its BYOD security implementation tactics. This highlights the increasing emphasis on risk identification and mitigation by the management.

This is exacerbated by the fact that the risks associated with BYOD are distinct in comparison to the traditional computing environment, necessitating the implementation of specific security measures [10, 47, 116]. Personal devices possess features that create specific risks to the organizations [11]. Due to ownership issues, device owners are responsible for securing their own devices, resulting in less security protection for devices compared to normal organization-owned computers. Personal devices also possess various applications that are susceptible to malicious codes. Studies also found mobile devices to be the most vulnerable endpoints to the organizational network and systems [90]. When devices are carried in and out of an organization, they may connect to unknown networks or be stolen or misplaced. The susceptibility of these personal devices allows malware to exploit those vulnerabilities.

The usage of personal devices that overlaps between the professional and personal life of employees weakens the security line and increases the risk of revealing organizational information. Inadequate malware detection leaves organizational data vulnerable to a variety of malicious parties. Moreover, the organization's security procedures would be unable to keep up with the constant, rapid evolution of mobile devices. Any attempt to manage employees' own devices appeared to compromise their freedom to use their own devices [56, 96]. Controlling these endpoints represent a violation of user privacy due to device ownership issues; hence, strategies that aided management in protecting traditional stationary computing systems may not be applicable in securing personal devices. As such, it is crucial to understand the threat landscape, identifying potential risks according to the workplace scenario and determine the right control measures to mitigate the security risks which will indirectly assist employees to comply with the BYOD security policies.

Organizations have information security policies as key instrument to reduce IS threats [60] however, it is widely reported that many employees fail to comply or deliberately ignored them [4, 62, 111, 113]. The purpose of ISP is to "give management direction and assistance for information security" [59]. A 2016 survey found that only 39% of BYOD-enabled firms had a policy in place, and 51% of BYOD employees have never read or seen their organization's policy [115]. Some employee may view the security policy as an inconvenience; hence, it will be disregarded. ISP is only effective if all employees adhere to it.

The adherence of an organization's information security policy (ISP) by its employees is crucial, as insider threats have a substantial impact on the credibility of the organization [114]. The vulnerabilities caused by the non-compliance conduct create opportunities for security incidents to occur [45]. According to studies, a significant proportion of security breaches result from noncompliance with security policies [50, 69, 111].

Failure to address these threats may cause data manipulations, modifications or destruction and denial of service [63]. Furthermore, for public sector organizations, the risks of information leakage are immense and may result in legal repercussions or loss of reputation. Additionally, it can detriment national security and sovereignty [38]. Hence, the practice of BYOD security behavior by public sector employees is crucial to reduce information security threats and improve information security management.

Although there is a burgeoning stream of research about information security policy compliance in organizations, to date, most studies focus on traditional computing environments where the desktops belong to the organization, leaving out the BYOD context. Studies exploring security compliance in a BYOD context are limited, notably in Malaysia. Protection Motivation Theory (PMT) was utilized in most studies, yet, the fact that there is one dominant theory in this field of study indicates that it is still a growing discipline [80, 86]. Most studies lack social and organizational factors, which is one of the limitations of extant research. Details of security policies are crucial as even though policies are in place, employees will still bypass these procedures if they are poorly designed [15, 66].

In response to this problem, this study aims to provide empirical findings to identify the factors that lead to employee BYOD security policy compliance and propose a BYOD security policy compliance framework to ensure the compliance of a security policy.

1.4 Research Objectives and Questions

As discussed in the previous section, BYOD has unique security risks that may require specific countermeasures, and insider threat is one of the organization's most significant challenges. In order to improve BYOD security in the Malaysian public sector, this study aims to examine the factors that influence BYOD security policy compliance among employees. Hence, the following research objectives (RO) and research questions (RQ) have been developed to address the research gaps as explained above:

RO1: To explore factors influencing BYOD security policy compliance intention.
RQ1: What are the factors that influence employees' compliance towards the BYOD security policy in an organization?

A review of human behavioral theories and models examining BYOD security policy compliance was conducted to identify plausible factors that formed the proposed BYOD security policy compliance research framework.

RO2: To identify the risks caused by employees' non-compliance behavior towards the BYOD security policy and control measures to mitigate the risks.

RQ2: What are the risks caused by employees' non-compliance behavior towards the BYOD security policy and what are the control measures to mitigate the risk?

This question seeks to explore the risks from BYOD security policy non-compliance behavior. Based on the findings, the risks are unique and the impact of such risks can be detrimental towards the security of the organizational information. Organizations need to understand how employees' behavior elevates risk and take steps to implement appropriate control measures to motivate employees to adhere to policies and procedures. In doing so, the processes that underpin compliance behaviors can be comprehended. Besides, one of the most critical components of security management is to identify and mitigate risks to reduce the risk occurrences [81]. These control measures can be expected to perform as risk reduction strategies as it enables employees to handle security threats better [22] and possibly influence employees' willingness to comply with the security policy [102]. The control measures contribute to the development of the proposed research framework as they are translated and mapped as factors that may influence employees' BYOD security policy compliance behavior.

To determine the impact of the factors and control measures on employees' BYOD security policy compliance, RO3 and RO4 were developed, as explained below.

RO3: To develop the BYOD security policy compliance framework.
RQ3: What is the relationship between the identified factors, control measures and employees' BYOD security policy compliance?
RO4: To validate the BYOD Security Policy Compliance Framework.
RQ4: How does the framework predict employees' compliance intention with the BYOD security policy?

This study investigates the impact of the factors on employees' compliance with the BYOD security policy in a CNII organizations. The proposed hypotheses were statistically validated to confirm the BYOD security policy compliance framework. This study employs predictive analysis to validate the framework's predictive capability. Research questions, objectives with relevant methods and expected outcomes outlined for this study is described in Table 1.1.

1.5 Scope of the Study

a. This research is centered on the public sector Critical National Information Infrastructure (CNII) agencies. The respondents targeted for this study were civil servants regardless of levels, grades, and schemes, who should adhere to the information security policies in the public sector organizations.
b. BYOD refers to employees bringing personal devices to the workplace and using such devices to connect to work-related systems. However, in this study, the BYOD concept is defined by devices such as laptops, tablets, and smartphones.

Table 1.1 Research objectives, questions and methods

No.	Research objectives	Research questions	Methods	Expected outcome
1	To explore factors influencing BYOD security policy compliance intention	What are the factors that influence employees' compliance towards the BYOD security policy in an organization?	• Systematic literature review 1	BYOD security policy compliance factors
2	To identify the risks caused by employees' non-compliance behavior towards BYOD security policy and control measures to mitigate the risk	What are the risks caused by employees' non-compliance behavior towards the BYOD security policy and what are the control measures to mitigate the risk?	• Systematic literature review 2 • Expert interview • Risks assessment	i. Risk factors ii. Control measures
3	To develop the BYOD security policy compliance framework	What is the relationship between the identified factors and employees' BYOD security policy compliance?	• Survey	BYOD security policy compliance framework
4	To validate the BYOD security policy compliance framework	How does the framework predict employees' compliance intention with the BYOD security policy?	• Coefficient of determi-nation, level of effect size and predictive relevance	Validated BYOD Security Policy Compliance Framework

c. The research employed Information Security Management System (ISMS) Risk Assessment Methodology (MyRAM) as a methodology to perform risk assessment in the qualitative phase.

1.6 Significance of Study

Given the widespread practice of BYOD in public sector enterprises, this research is necessary from the researcher's perspective. Identification of factors that influence employees' intention toward complying with BYOD security policy would be valuable in achieving information security objectives [29]. The arguments supporting the significance of this research are discussed from both theoretical and practical perspectives.

a. **Theoretical Contribution**

Personal devices present a plethora of risks that require special consideration. Existing models and theories must be reexamined in light of BYOD phenomenon.

The work makes a substantial contribution to the burgeoning field of BYOD security policy compliance research and identifies avenues for future investigation.

This research will contribute to the body of knowledge in the field of information security policy compliance in the context of BYOD. As the security concerns posed by BYOD in an organization are significantly different from those of a conventional computing environment, compliance behavior of employees in a BYOD environment is likely to be influenced by a variety of circumstances. Generalization of existing theories and models may not be applicable, or the models' components may vary considerably in predicting BYOD security policy compliance in the public sector. Combined with the complexity of human behavior, the necessity to conduct the research is established. This study utilizes control measures that mitigate risks and underexplored factors to identify determinants of BYOD security policy compliance. The control measures and factors are interpreted into relevant human behavioral theories to determine the association between the identified factors and BYOD security policy compliance among selected civil servants.

This study is significant as it employs mixed method. Firstly, instead of focusing only on the exploratory findings from qualitative technique alone, which might be a biased viewpoint, the study confirms the findings with quantitative approach. The proposed research framework is formed from the findings of systematic literature review and the insights of qualitative (semi-structured interviews and risks assessment with security experts). Subsequently, the conceptualized framework is tested quantitatively (survey distributed to employees) to obtain consensus on the factors influencing BYOD security policy compliance. Thus, the BYOD security policy framework will be tested efficiently to understand public sector employees' security compliance behavior.

b. **Practical Contribution**

This study is significant as it highlights the factors that contributes to the compliancy of BYOD security policy. Knowledge and understanding of these factors may provide additional insights to improve BYOD security policy implementation. The findings from this study could inform the Information Technology (IT) management on the security policy compliance when implementing a BYOD program and hence improve their information security management.

As we all know, the BYOD trend is here to stay and is expanding with the current 4 billion mobile subscribers worldwide. Even if organizations spend millions of dollars enhancing their security technology, it will be ineffective if employees are unwilling to comply with the BYOD security policy. As such, this research attempts to provide insights that contribute to a better understanding of human behavior regarding BYOD ISP compliance. The research findings will enhance the most important pillar of a business, the 'People,' by supporting policymakers in developing BYOD security policies that are likely to be adhered to by employees.

Fig. 1.1 Main stages of the research process

1.7 Overview of Research Methodology

The current study adopted mixed method research approach in which qualitative and quantitative techniques were executed sequentially and examined at various phases of the research process within a single study. The research methodology is described in detail in Chap. 4.

The research comprised of three main phases (as depicted in Fig. 1.1). Initially, the research problem, research questions, objectives, and contributions are identified. In the first phase, Systematic Literature Review (SLR) was performed in order to explore the factors that contribute to BYOD security policy compliance. This also allowed for the identification of gaps in the literature. The findings were used to construct proposed research framework and research hypotheses.

In the second phase, a SLR was carried out to identify risk factors. Next, using these findings as a foundation, expert interviews and risks assessment were conducted to gather insight into the research problem, delve deeper into the BYOD risks, and identify control measures that mitigate risk which were mapped into the conceptual framework.

Phase three commenced in June 2020 through May 2021, over a period of eleven months involving a self-administered online survey in three CNII agencies. A total of 346 valid responses were gathered and the data were analyzed using Partial Least Squares—Structural Equation Modelling (PLS-SEM). The statistical analysis enabled the structural model analysis to confirm the research's proposed hypotheses and the model's predictive ability.

1.8 Organization of Book

The book is divided into six chapters. The six chapters of the study are organized as follows:

This chapter depicts the study's context, problem statement, definitions, and terms utilized in the study in the research. The research aims and questions of the current study research are also identified and justified in this chapter. The research objectives were developed in response to the insecure BYOD practices in the public sector, which are harmful to organizational information security. In this chapter, the significance of the research, the scope of the research, a brief description of the research methodology, and the overall structure of the book are all presented.

Chapter 2 examines the BYOD practice, benefits, BYOD security concerns, mitigation solutions, and variables impacting BYOD security policy compliance. This chapter reviews and describes some theories and models linked to human behavior and information security. Significant factors in BYOD security policy compliance are explored on its impact on compliance behavior. Reviewing provides vital insights for developing the BYOD security policy compliance framework and proposing hypotheses for this study. This chapter concludes by highlighting literature gaps.

Chapter 3 presents the theoretical foundations and the proposed conceptual framework for this study. The explanation of each factor proposed in the framework and hypotheses of possible associations is presented.

Chapter 4 begins with a discussion of research paradigms and a description of research design, which encompasses the technique employed in the current study, including the justification for selecting mixed method research. This chapter also discusses the conceptualization of the BYOD security policy compliance framework, which led to the formulation of causal links (hypotheses) in the research framework. The mixed method technique used is described in detail, including qualitative (expert interviews) and quantitative research (self-administered survey). It also describes the research methodology, including the search strategy, sample method, and procedures for generating the survey instrument used in the current study.

Chapter 5 presents the qualitative and quantitative results from the security experts and survey. Findings on BYOD risks, risk severity and control measures that mitigate risks were presented and discussed as outcome of the qualitative research. This chapter also presents a detailed account of the findings of the quantitative study. The results were analyzed using IBM SPSS Statistics 24 and Smart PLS version 3.3. The findings of the descriptive analysis were provided first, followed by PLS-SEM analysis, which highlights the significant factors that lead to BYOD security policy compliance. A discussion of the PLS SEM analysis for each discovered factor is provided.

Chapter 6 is devoted to an overall summary of the research and discusses the achievement of the research aim and objectives. The results from the qualitative and quantitative analysis are discussed in line with the research objectives, followed by the presentation of the research contributions and limitations. This chapter concludes with recommendations for future study and a research conclusion.

References

1. Abdul Molok, N.N., Ahmad, A., Chang, S.: Exploring the use of online social networking by employees: looking at the potential for information leakage. In: Pacific Asia Conference on Information Systems—PACIS, p. 138 (2011)
2. Abdul Molok, N.N., Ahmad, A., Chang, S.: A case analysis of securing organisations against information leakage through online social networking. Int. J. Inf. Manag. **43**, 351–356 (2018). https://doi.org/10.1016/j.ijinfomgt.2018.08.013
3. Agudelo-Serna, C.A., Ahmad, A., Bosua, R., Maynard, S.B.: Strategies to mitigate knowledge leakage risk caused by the use of mobile devices: a preliminary study. In: 38th International Conference on Information Systems (ICIS 2017), pp. 1–24 (2017)
4. Agudelo-Serna, C., Bosua, R., Ahmad, A., Maynard, S.B.: Towards a knowledge leakage mitigation framework for mobile devices in knowledge-intensive organizations. In: 26th European Conference on Information Systems: Beyond Digitization—Facets of Socio-Technical Change, ECIS 2018 (2018)
5. Agudelo, C.A., Bosua, R., Ahmad, A., Maynard, S.B.: Mitigating knowledge leakage risk in organizations through mobile devices: a contextual approach. In: 27th Australasian Conference on Information Systems 2016—ACIS2016, pp. 1–12 (2016)
6. Alassaf, M., Alkhalifah, A.: Exploring the influence of direct and indirect factors on information security policy compliance: a systematic literature review. IEEE Access **9**, 162687–162705 (2021). https://doi.org/10.1109/access.2021.3132574
7. AlGhamdi, S., Win, K.T., Vlahu-Gjorgievska, E.: Employees' intentions toward complying with information security controls in Saudi Arabia's public organisations. Gov. Inf. Q. 101721 (2022). https://doi.org/10.1016/j.giq.2022.101721
8. Almarhabi, K., Bahaddad, A., Mohammed, A.: Security management of BYOD and cloud environment in Saudi Arabia. Alex. Eng. J. (2022). https://doi.org/10.1016/j.aej.2022.07.031
9. Ameen, N., Hosany, S., Tarhini, A.: Consumer interaction with cutting-edge technologies: trends and implications for future research. Comput. Hum. Behav. 135907 (2021). https://doi.org/10.1016/j.chb.2021.106761
10. Ameen, N., Tarhini, A., Hussain Shah, M., Madichie, N.O.: Employees' behavioural intention to smartphone security: a gender-based, cross-national study. Comput. Hum. Behav. **104**(106184) (2020). https://doi.org/10.1016/j.chb.2019.106184
11. Ameen, N., Tarhini, A., Shah, M.H., Madichie, N., Paul, J., Choudrie, J.: Keeping customers' data secure: a cross-cultural study of cybersecurity compliance among the Gen-Mobile workforce. Comput. Hum. Behav.**114** (2021). https://doi.org/10.1016/j.chb.2020.106531
12. Amran, N.A., Ahmad, H.N., Hassan, N.L.: Malaysian public sector size: a comparison with other ASEAN countries. J. Bus. Manag. Account. **11**, 1–20 (2021). https://doi.org/10.32890/jbma2021.11.1.1
13. Balozian, P., Leidner, D.: Review of IS security policy compliance. ACM SIGMIS Database **48**(3), 11–43 (2017). https://doi.org/10.1145/3130515.3130518
14. Bautista, J.R., Rosenthal, S., Lin, T.T.C., Theng, Y.L.: Predictors and outcomes of nurses' use of smartphones for work purposes. Comput. Hum. Behav. **84**, 360–374 (2018). https://doi.org/10.1016/j.chb.2018.03.008
15. Beautement, A., Sasse, M.A., Wonham, M.: The compliance budget: managing security behaviour in organisations. In: Proceedings of the 2008 New Security Paradigms Workshop, pp. 47–58 (2008)
16. Bloomberg: Malaysian military data stolen by suspected state hackers last year. Free Malaysia Today (2023). https://www.freemalaysiatoday.com/category/nation/2023/01/11/malaysian-military-data-stolen-from-suspected-state-hackers-last-year/
17. Bulgurcu, B., Cavusoglu, H., Benbasat, I.: Information security policy compliance: an empirical study of rationality-based beliefs and information security awareness. MIS Q. **34**(3), 523–548 (2010). https://doi.org/10.1093/bja/aeq366

18. Burns, A.J., Roberts, T.L., Posey, C., Lowry, P.B., Fuller, B.: Going beyond deterrence: a middle-range theory of motives and controls for insider computer abuse. Inf. Syst. Manag. (2022)
19. Butavicius, M., Parsons, K., Lillie, M., Mccormac, A., Pattinson, M., Calic, D.: When believing in technology leads to poor cyber security: development of a trust in technical controls scale. Comput. Secur. 102020 (2020). https://doi.org/10.1016/j.cose.2020.102020
20. Cappelli, D.M., Moore, A.P., Trzeciak, R.F.: The CERT Guide to Insider Threats: How to Prevent, Detect, and Respond to Information Technology Crimes (Theft, Sabotage, Fraud). Addison-Wesley Professional (2012)
21. Chen, X., Wu, D., Chen, L., Teng, J.K.L.: Sanction severity and employees' information security policy compliance: investigating mediating, moderating, and control variables. Inf. Manag. **55**(8), 1049–1060 (2018). https://doi.org/10.1016/j.im.2018.05.011
22. Chen, Y., Galletta, D.F., Lowry, P.B., Luo, X. (Robert), Moody, G.D., Willison, R.: Understanding inconsistent employee compliance with information security policies through the lens of the extended parallel process model. Inf. Syst. Res. (2021). https://doi.org/10.1287/isre.2021.1014
23. Cho, V., Ip, W.H.: A study of BYOD adoption from the lens of threat and coping appraisal of its security policy. Enterp. Inf. Syst. **12**(6), 659–673 (2018). https://doi.org/10.1080/17517575.2017.1404132
24. Chu, A.M.Y., Chau, P.Y.K.: Development and validation of instruments of information security deviant behavior. Decis. Support. Syst. **66**, 93–101 (2014). https://doi.org/10.1016/j.dss.2014.06.008
25. Chua, H.N., Teh, J.S., Herbland, A.: Identifying the effect of data breach publicity on information security awareness using hierarchical regression. IEEE AccessPP, 1–1 (2021). https://doi.org/10.1109/access.2021.3107426
26. Costa, D.: CERT Definition of "Insider Threat"—Updated. Software Engineering Institute, Carnegie Mellon University (2021). https://insights.sei.cmu.edu/blog/cert-definition-of-insider-threat-updated/
27. CPNI: Cyber Insider (2020). https://www.cpni.gov.uk/insider-threat
28. Cram, W.A., Proudfoot, J.G., D'Arcy, J.: Organizational information security policies: a review and research framework. Eur. J. Inf. Syst. **26**(6), 605–641 (2017). https://doi.org/10.1057/s41303-017-0059-9
29. Cram, W.A., Proudfoot, J.G., D'Arcy, J.: Maximizing employee compliance with cybersecurity policies. MIS Q. Exec. **19**(3), 183–198 (2020). https://doi.org/10.17705/2msqe.00032
30. Cyber Security Malaysia: National Cyber Security Policy (NCSP) (2021). https://cnii.cybersecurity.my/main/ncsp/tncsp.html
31. Dalal, R.S., Howard, D.J., Brummel, B.J., Bennett, R.J.: Organizational science and cybersecurity: abundant opportunities for research at the interface (2021)
32. Dang-Pham, D., Kautz, K., Hoang, A.-P., Pittayachawan, S.: Identifying information security opinion leaders in organizations: insights from the theory of social power bases and social network analysis. Comput. Secur. **112**, 102505 (2022). https://doi.org/10.1016/j.cose.2021.102505
33. Daud, M.: National Information Infrastructure Organisations and Cyber Security Compliance in Malaysia. Doctoral thesis, University of Malaya (2018)
34. Davison, R.M.: The transformative potential of disruptions: a viewpoint. Int. J. Inf. Manag. **55**(102149), 1–4 (2020). https://doi.org/10.1016/j.ijinfomgt.2020.102149
35. De', R., Pandey, N., Pal, A.: Impact of digital surge during Covid-19 pandemic: a viewpoint on research and practice. Int. J. Inf. Manag.**55**, 102171 (2020). https://doi.org/10.1016/j.ijinfomgt.2020.102171
36. Dhillon, G., Smith, K., Dissanayaka, I.: Information systems security research agenda: exploring the gap between research and practice. J. Strateg. Inf. Syst. **30**(4), 101693 (2021). https://doi.org/10.1016/j.jsis.2021.101693

37. Donalds, C., Osei-Bryson, K.M.: Cybersecurity compliance behavior: exploring the influences of individual decision style and other antecedents. Int. J. Inf. Manag. 102056 (2019). https://doi.org/10.1016/j.ijinfomgt.2019.102056
38. Dzazali, S., Sulaiman, A., Zolait, A.H.: Information security landscape and maturity level: case study of Malaysian Public Service (MPS) organizations. Gov. Inf. Q. **26**(4), 584–593 (2009). https://doi.org/10.1016/j.giq.2009.04.004
39. Eastin, M.S., Brinson, N.H., Doorey, A., Wilcox, G.: Living in a big data world: predicting mobile commerce activity through privacy concerns. Comput. Hum. Behav. **58**, 214–220 (2016). https://doi.org/10.1016/j.chb.2015.12.050
40. Ekran: What Is an Insider Threat? Definition, Types, and Countermeasures. Ekran (2021). https://www.ekransystem.com/en/blog/insider-threat-definition
41. Engan, J.: Cabaran BYOD dalam persekitaran kerja [PowerPoint]. Dokumen Site (2018)
42. Eom, S.-J., Lee, J.: Digital government transformation in turbulent times: responses, challenges, and future direction. Gov. Inf. Q. 101690 (2022). https://doi.org/10.1016/j.giq.2022.101690
43. French, A., Guo, C., Shim, J.P.: Current status, issues, and future of bring your own device (BYOD). Commun. Assoc. Inf. Syst. **35**(10) (2014)
44. Furnell, S., Clarke, N.: Power to the people? The evolving recognition of human aspects of security. Comput. Secur. **31**(8), 983–988 (2012). https://doi.org/10.1016/j.cose.2012.08.004
45. Gangire, Y., Da Veiga, A., Herselman, M.: Assessing information security behaviour: a self-determination theory perspective. Inf. Comput. Secur. (2021). https://doi.org/10.1108/ICS-11-2020-0179
46. Geoff, E.: Bring Your Own Device—The New Normal: The NCSC View on BYOD and the Rise in Home Working [Blog]. National Cyber Security Centre (2020)
47. Giwah, A.D., Wang, L., Levy, Y., Hur, I.: Empirical assessment of mobile device users' information security behavior towards data breach—leveraging protection motivation theory. J. Intellect. Cap. **21**(2), 215–233 (2019). https://doi.org/10.1108/JIC-03-2019-0063
48. Gnaneswaran, D.: Cybersecurity threats to cost organizations in Malaysia US$12.2 billion in economic losses (2018). https://news.microsoft.com/en-my/2018/07/12/cybersecurity-threats-to-cost-organizations-in-malaysia-us12-2-billion-in-economic-losses/
49. Grover, V.: BYOD Trends in 2019: Its Influence in the Next Five Years (2019). https://blog.scalefusion.com/top-8-trends-that-will-influence-byod-in-next-five-years/
50. Gwebu, K.L., Wang, J., Hu, M.Y.: Information security policy noncompliance: an integrative social influence model. Inf. Syst. J. **30**(2), 1–50 (2019). https://doi.org/10.1111/isj.12257
51. Harris, J., Ives, L., Junglas, I.: IT consumerization: when gadgets turn into enterprise IT tools. MIS Q. Exec. **11**(3), 99–112 (2012)
52. Harris, M.A., Patten, K., Regan, E.: The need for BYOD mobile device security awareness and training. In: 19th Americas Conference on Information Systems (AMCIS 2013) (2013)
53. Hoh, G.C.: Head in the Clouds: how remote working behaviours are exposing organisations to cyber risks. Business Today (2020)
54. Holtsnider, B., Jaffe, B.D.: IT Manager's Handbook: Getting Your New Job Done. Elsevier Inc. (2012). https://doi.org/10.1016/b978-012370488-7/50001-4
55. Hong, Y., Furnell, S.: Understanding cybersecurity behavioral habits: insights from situational support. J. Inf. Secur. Appl. **57**, 102710 (2021). https://doi.org/10.1016/j.jisa.2020.102710
56. Hovav, A., Putri, F.F.: This is my device! Why should I follow your rules? Employees' compliance with BYOD security policy. Pervasive Mob. Comput. **32**, 35–49 (2016). https://doi.org/10.1016/j.pmcj.2016.06.007
57. Hwang, K., Choi, M.: Effects of innovation-supportive culture and organizational citizenship behavior on e-government information system security stemming from mimetic isomorphism. Gov. Inf. Q. **34**(2), 183–198 (2017). https://doi.org/10.1016/j.giq.2017.02.001
58. Ifinedo, P.: Information systems security policy compliance: an empirical study of the effects of socialisation, influence, and cognition. Inf. Manag. **51**(1), 69–79 (2014). https://doi.org/10.1016/j.im.2013.10.001

59. International Organization for Standardization: ISO/IEC 27001:2013—Information technology—Security techniques—Information security management systems—Requirements. ISO (2013)
60. Jaeger, L., Eckhardt, A., Kroenung, J.: The role of deterrability for the effect of multi-level sanctions on information security policy compliance: results of a multi-group analysis. Inf. Manag.minor revision, 103318 (2020). https://doi.org/10.1016/j.im.2020.103318
61. Jarrahi, M.H., Crowston, K., Bondar, K., Katzy, B.: A pragmatic approach to managing enterprise IT infrastructures in the era of consumerization and individualization of IT. Int. J. Inf. Manag. **37**(6), 566–575 (2017). https://doi.org/10.1016/j.ijinfomgt.2017.05.016
62. Johnston, A.C., Warkentin, M., Dennis, A.R., Siponen, M.: Speak their language: designing effective messages to improve employees' information security decision making. Decis. Sci. **50**(2), 245–284 (2019). https://doi.org/10.1111/deci.12328
63. Kearns, G.: Countering mobile device threats: a mobile device security model. J. Forens. Invest. Account. **8**(1), 157–167 (2016). http://www.nacva.com/jfia
64. Khan, N.F., Yaqoob, A., Khan, M.S., Ikram, N.: The cybersecurity behavioral research: a tertiary study. Comput. Secur. 102826 (2022). https://doi.org/10.1016/j.cose.2022.102826
65. Kielt, C.: Tech companies need to sell the way digital natives want to buy. Reltio (2020). https://techcrunch.com/sponsor/reltio/tech-companies-need-to-sell-the-way-digital-natives-want-to-buy/
66. Kirlappos, I., Parkin, S., Sasse, M.A.: "Shadow security" as a tool for the learning organization. ACM SIGCAS Comput. Soc. **45**(1), 29–37 (2015). https://doi.org/10.1145/2738210.2738216
67. Koohang, A., Anderson, J., Nord, J.H., Paliszkiewicz, J.: Building an awareness-centered information security policy compliance model. Ind. Manag. Data Syst. **120**(1), 231–247 (2019). https://doi.org/10.1108/IMDS-07-2019-0412
68. Lallie, H.S., Shepherd, L.A., Nurse, J.R.C., Erola, A., Epiphaniou, G., Maple, C., Bellekens, X.: Cyber security in the age of COVID-19: a timeline and analysis of cyber-crime and cyber-attacks during the pandemic. Comput. Secur. **105**, 102248 (2021). https://doi.org/10.1016/j.cose.2021.102248
69. Lowry, P.B., Moody, G.D.: Proposing the control-reactance compliance model (CRCM) to explain opposing motivations to comply with organisational information security policies. Inf. Syst. J. **25**(5), 433–463 (2015). https://doi.org/10.1111/isj.12043
70. Lowry, P.B., Posey, C., Roberts, T.L., Bennett, R.J.: Is your banker leaking your personal information? The roles of ethics and individual-level cultural characteristics in predicting organizational computer abuse. J. Bus. Ethics **121**(3), 385–401 (2014). https://doi.org/10.1007/s10551-013-1705-3
71. Lugo, J.: Bring Your Own Device (BYOD): Bitglass' 2020 personal device report. Bitglass (2020)
72. MarketWatch: Bring-your-own-Device (BYOD) Market Size, Share, Industry, Analysis, Price, Trends, Growth, Report and Forecast 2020–2025 (2021). https://www.marketwatch.com/press-release/bring-your-own-device-byod-market-size-share-industry-analysis-price-trends-growth-report-and-forecast-2020-2025-2021-04-12?tesla=y
73. Mat Akhir, H.: Success model for BYOD implementation considering human and security perspectives in Malaysia Government environment (2017)
74. Mehra, A., Rajput, S., Paul, J.: Determinants of adoption of latest version smartphones: theory and evidence. Technol. Forecast. Soc. Change **175**, 121410 (2022). https://doi.org/10.1016/j.techfore.2021.121410
75. Merhi, M.I., Ahluwalia, P.: Examining the impact of deterrence factors and norms on resistance to Information Systems Security. Comput. Hum. Behav. **92**, 37–46 (2019). https://doi.org/10.1016/j.chb.2018.10.031
76. Moody, G.D., Siponen, M., Pahnila, S.: Toward a unified model of information security policy compliance. MIS Q. **42**(1), 285–311 (2018). https://doi.org/10.25300/MISQ/2018/13853
77. Mueller, M., Klesel, M., Heger, O., Niehaves, B.: Empirical insights on individual innovation behaviour: a qualitative study on it—consumerization. In: PACIS Proceedings, p. 201 (2016)

78. MyDigital: Malaysia Digital Economy Blueprint. Economic Planning Unit Prime Minister's Department, p. 104 (2021)
79. National Security Council: Malaysia Cyber Security Strategy 2020–2024, pp. 1–90 (2020)
80. Nehme, A., Warkentin, M., Jang, K., Kim, S.: Information Security and Privacy Beyond Rational Information Security Decisions: An Alternate Beyond Rational Information Security Decisions: An Alternate View Abstract. AMCIS 2022 SIG SEC—Information Security and Privacy, 0–5 (2022)
81. Ng, B.Y., Kankanhalli, A., Xu, Y. (Calvin): Studying users' computer security behavior: a health belief perspective. Decision Support Syst. **46**(4), 815–825 (2009). https://doi.org/10.1016/j.dss.2008.11.010
82. Niehaves, B., Köffer, S., Ortbach, K., Reimler, S.: Boon and bane of IT consumerization: the burnout-engagement-continuum. In: 19th Americas Conference on Information Systems (AMCIS 2013), pp. 1–9 (2013). http://aisel.aisnet.org/amcis2013/EndUserIS/GeneralPresentations/11/
83. Ogbanufe, O., Pavur, R.: Going through the emotions of regret and fear: revisiting protection motivation for identity theft protection. Int. J. Inf. Manag. **62**, 102432 (2022). https://doi.org/10.1016/j.ijinfomgt.2021.102432
84. Oracle, & KPMG: Oracle and KPMG cloud threat report 2019—defining edge intelligence: closing visibility gaps with a layered defense strategy read full report (2019)
85. Paananen, H., Lapke, M., Siponen, M.: State of the art in information security policy development. Comput. Secur. **88**, 101608 (2020). https://doi.org/10.1016/j.cose.2019.101608
86. Palanisamy, R., Norman, A.A., Kiah, M.L.M.: Compliance with Bring Your Own Device security policies in organizations: a systematic literature review. Comput. Secur. 101998 (2020). https://doi.org/10.1016/j.cose.2020.101998
87. Papagiannidis, S., Harris, J., Morton, D.: WHO led the digital transformation of your company? A reflection of IT related challenges during the pandemic. Int. J. Inf. Manag. **55**(102166), 1–5 (2020). https://doi.org/10.1016/j.ijinfomgt.2020.102166
88. Parsons, K., Calic, D., Pattinson, M., Butavicius, M., McCormac, A., Zwaans, T.: The Human Aspects of Information Security Questionnaire (HAIS-Q): two further validation studies. Comput. Secur. **66**, 40–51 (2017). https://doi.org/10.1016/j.cose.2017.01.004
89. Perimbanayagam, K., Nabila, D.: Personal info of 800,000 voters compromised by alleged breach of EC database. New Straits Times (2022). https://www.nst.com.my/news/crime-courts/2022/11/849700/personal-info-800000-voters-compromised-alleged-breach-ec-database
90. Ponemon Institute: 2018 State of Cybersecurity in Small & Medium Size Businesses, p. 46 (2018). https://www.keepersecurity.com/assets/pdf/Keeper-2018-Ponemon-Report.pdf
91. Ponemon Institute: 2020 Cost of Insider Threats Global Report (2020)
92. Posey, C., Raja, U., Crossler, R.E., Burns, A.J.: Taking stock of organisations' protection of privacy: categorising and assessing threats to personally identifiable information in the USA. Eur. J. Inf. Syst. **26**(6), 585–604 (2017). https://doi.org/10.1057/s41303-017-0065-y
93. Posey, C., Roberts, T.L., Lowry, P.B., Bennett, R.J., Courtney, J.F.: Insiders' protection of organizational information assets: development of a systematics-based taxonomy and Theory Of Diversity for Protection-Motivated behaviors. MIS Q. **37**(4), 1189–1210 (2013)
94. Posey, C., Roberts, T.L., Lowry, P.B., Hightower, R.T.: Bridging the divide: a qualitative comparison of information security thought patterns between information security professionals and ordinary organizational insiders. Inf. Manag. **51**(5), 551–567 (2014). https://doi.org/10.1016/j.im.2014.03.009
95. Prime Minister's Office: Speech text of cyber defence & security exhibition and conference 2021. In: Cyber Defence & Security Exhibition and Conference (CYDES) 2021 (2021). https://www.pmo.gov.my/2021/06/speech-text-of-cyber-defence-security-exhibition-and-conference-2021/
96. Putri, F., Hovav, A.: Employees' compliance with BYOD security policy: insights from reactance, organizational justice, and Protection Motivation Theory. In: 22nd European Conference on Information Systems (ECIS 2014), pp. 1–17 (2014). https://doi.org/10.1109/ISCAIE.2014.7010235

97. Rajab, M., Eydgahi, A.: Evaluating the explanatory power of theoretical frameworks on intention to comply with information security policies in higher education. Comput. Secur. **80**, 211–223 (2019). https://doi.org/10.1016/j.cose.2018.09.016
98. Rivera, A.D., George, G., Peter, P., Muralidharan, S., Khanum, S.: Analysis of Security Controls for BYOD (Bring Your Own Device), pp. 1–10 (2013)
99. Safa, N.S., Maple, C.: Human errors in the information security realm—and how to fix them. Comput. Fraud Secur. **2016**(9), 17–20 (2016). https://doi.org/10.1016/S1361-3723(16)30073-2
100. Safa, N.S., Maple, C., Furnell, S., Azad, M.A., Perera, C., Dabbagh, M., Sookhak, M.: Deterrence and prevention-based model to mitigate information security insider threats in organisations. Futur. Gener. Comput. Syst. (2019). https://doi.org/10.1016/j.future.2019.03.024
101. Safa, N.S., Von Solms, R., Furnell, S.: Information security policy compliance model in organizations. Comput. Secur. **56**, 1–13 (2016). https://doi.org/10.1016/j.cose.2015.10.006
102. Sanyal, P., Menon, N., Siponen, M.: An empirical examination of the economics of mobile application security. MIS Q. **45**(4), 2235–2260 (2021). https://doi.org/10.25300/MISQ/2021/15315
103. Shammugam, I.: Information security risk assessment using survival analysis technique for the Malaysian Public Sector data centers [Universiti Teknologi Malaysia] (2021). https://doi.org/10.3390/encyclopedia1030050
104. Shin, B., Lowry, P.B.: A review and theoretical explanation of the 'Cyberthreat-Intelligence (CTI) capability' that needs to be fostered in information security practitioners and how this can be accomplished. Comput. Secur. **92**, 101761 (2020). https://doi.org/10.1016/j.cose.2020.101761
105. Siponen, M.T., Mahmood, M.A., Pahnila, S.: Employees' adherence to information security policies: an exploratory field study. Inf. Manag. **51**(2), 217–224 (2014). https://doi.org/10.1016/j.im.2013.08.006
106. Stagliano, T., Dipoalo, A., Coonelly, P.: Consumerization of IT. Mathematics and Computer Science Capstones, p. 10 (2013). https://digitalcommons.lasalle.edu/mathcompcapstones/10
107. Statista: Number of smartphone subscriptions worldwide from 2016 to 2027 (2022). https://www.statista.com/statistics/330695/number-of-smartphone-users-worldwide/https://www.statista.com/statistics/330695/number-of-smartphone-users-worldwide/
108. Stewart, H., Jürjens, J.: Information security management and the human aspect in organizations. Inf. Comput. Secur. **25**(5), 494–534 (2017). https://doi.org/10.1108/ICS-07-2016-0054
109. Straub, D.W., Welke, R.J.: Coping with systems risk: security planning models for management decision making. MIS Q. **22**(4), 441 (1998). https://doi.org/10.2307/249551
110. Syed Soffian, S.I., Engku Ismail, E.A., Faudziah Hanim, H.F., Hassan, H.: Public sector accounting and financial management in Malaysia. Pearson Malaysia Sdn Bhd. (2011)
111. Tam, C., De Matos, C., Oliveira, T.: What influences employees to follow security policies? Saf. Sci. **147** (2022). https://doi.org/10.1016/j.ssci.2021.105595
112. Tariq, Q.: Online dangers during the pandemic. The Star Malaysia, 14 Sept (2020). https://www.pressreader.com/malaysia/the-star-malaysia-star2/20200914/281913070528015
113. Timms, K.: BYOD must be met with a wider appreciation of the cyber-security threat. Comput. Fraud Secur. **2017**(7), 5–8 (2017). https://doi.org/10.1016/S1361-3723(17)30058-1
114. Trinkle, B.S., Warkentin, M., Malimage, K., Raddatz, N.: High-risk deviant decisions: does neutralization still play a role? J. Assoc. Inf. Syst. **22**(3), 797–826 (2021). https://doi.org/10.17705/1jais.00680
115. Trustlook: Trustlook Insights Q4 report (2016). https://newblogtrustlook.files.wordpress.com/2016/10/trustlook_insights_q4_2016_byod.pdf
116. Tu, Z., Turel, O., Yuan, Y., Archer, N.: Learning to cope with information security risks regarding mobile device loss or theft: an empirical examination. Inf. Manag. **52**(4), 506–517 (2015). https://doi.org/10.1016/j.im.2015.03.002

117. Vallejo-Correa, P., Monsalve-Pulido, J., Tabares-Betancur, M.: Systematic mapping review of context-aware analysis and its approach to mobile learning and ubiquitous learning processes. Comput. Sci. Rev.**39** (2021). https://doi.org/10.1016/j.cosrev.2020.100335
118. Wani, T.A., Mendoza, A., Gray, K.: BYOD usage and security behaviour of hospital clinical staff: an Australian survey. Int. J. Med. Inform. 104839 (2022). https://doi.org/10.1016/j.ijmedinf.2022.104839
119. Weeger, A., Wang, X., Gewald, H., Raisinghani, M., Otavio, S., Grant, G., Siddhi, P.: Determinants of intention to participate in corporate BYOD-programs: the case of digital natives. Inf. Syst. Front. 1–42 (2018). https://doi.org/10.13140/RG.2.1.5160.0726
120. Wilder, C.R.: IT & Security Operations Teams Must Be Ready for 2021 and Beyond. Forbes (2021). https://www.forbes.com/sites/moorinsights/2021/04/15/it--security-operations-teams-must-be-ready-for-2021-and-beyond/
121. Wiley, A., McCormac, A., Calic, D.: More than the individual: examining the relationship between culture and information security awareness. Comput. Secur.**88** (2020). https://doi.org/10.1016/j.cose.2019.101640
122. Willison, R., Warkentin, M.: Beyond deterrence: an expanded view of employee computer abuse. MIS Q. **37**(1), 1–20 (2013). https://doi.org/10.1080/01639625.2012.759048
123. Wong, L., Lee, V., Tan, G.W., Ooi, K.: The role of cybersecurity and policy awareness in shifting employee compliance attitudes: building supply chain capabilities. Int. J. Inf. Manag. **66**, 102520 (2022). https://doi.org/10.1016/j.ijinfomgt.2022.102520
124. Yunos, Z., Ahmad, R., Suid, S.H., Ismail, Z.: Safeguarding Malaysia's Critical National Information Infrastructure (CNII) against cyber terrorism: towards development of a policy framework. In: 2010 6th International Conference on Information Assurance and Security, IAS 2010, pp. 21–27 (2010). https://doi.org/10.1109/ISIAS.2010.5604182

Chapter 2
Bring Your Own Device

The first chapter explained the research's context, topic, objectives, questions, and its theoretical and practical contributions. This chapter begins with a literature analysis of BYOD characteristics, benefits and security risks and BYOD security policy compliance models and theories. The aim of thorough review is to lay the foundation for the discovery of vital factors influencing BYOD security policy compliance. The reviews also support in illustrating the gaps in the existing body of knowledge and rationalize the purpose of conducting this research.

2.1 Bring Your Own Device (BYOD)

BYOD, which refers to the practice of employees bringing their own devices to the workplace, has become the standard in recent years as a result of the growing adoption of mobile devices such as smartphones and tablets [137]. In certain studies, BYOD is also classified as IT consumerization [206], due to the expanding role and impact that consumer-generated technologies have on enterprises [102]. The BYOD market size is estimated to hit around USD 430.25 billion by the year of 2025 as compared to USD 186.09 billion in 2019 [144]. It is a preference among employees especially the younger generation to use their own devices without being controlled by their organizations [98, 164].

The technology trend has shifted from traditional top-down IT innovation to a consumer-driven bottom-up strategy [128, 230], prompted by employees who are driving privately used IT solutions into the workplace [233].

The use of mobile personal devices, such as laptop computers, smartphones, and tablets, is an integral element of the workday [225]. BYOD has become so ingrained in employees' personal and professional life that it has turned into a bottom-up culture in enterprises, controlling how people work and what devices they use [20, 128, 129]. An increasing number of digital natives, the new generation of employees

© The Author(s), under exclusive license to Springer Nature Switzerland AG 2025

R. Palanisamy et al., *Bring Your Own Device Security Policy Compliance Framework*, Information Systems Engineering and Management 37,
https://doi.org/10.1007/978-3-031-86801-6_2

who have grown up with smartphones [112], will enter the workforce and demand the right to use their personal hardware and software at work [231].

2.2 BYOD Features

BYOD comes with a set of significant features that serve as distinct advantages and unique drawbacks to the organizations and it is detailed in the following subsection.

2.2.1 *Mobility*

The exponential rise in popularity of bring-your-own-device over the past decade has revolutionized the method in which people carry out their work duties. With the proliferation of devices, employees carry their personal devices to work and their homes while accessing corporate emails from anywhere on their devices as long as Wi-Fi or mobile data are available [3]. Employees use personal devices to communicate, learning, collaborating and getting things done while working remotely [19, 225]. BYOD increases ubiquitous communication between employees and managers by eliminating accessibility issues in terms of physical locations and time. Employees, especially digital natives, are fast to adopt BYOD [86, 92], as they require quick responses, increased access to technology, workplace flexibility and continuous interaction. Employees remain connected continuously to the digital world via their personal devices while blurring the lines between the physical office and the place where work takes place [60]. Formerly, employees were stationed and protected within a well-defined security perimeter, but this setting no longer suits their changing needs [28]. It is projected that BYOD will increase the number of employees who work remotely.

2.2.2 *Device Ownership*

BYOD has shifted the element of ownership to the employees, which distinguishes it from organization-owned devices [3]. The attachment of personal devices to the behavior and preferences of their users has transformed the devices into an extension of the individual [86]. There is a sense of connectedness because mobile devices are also utilized to communicate with family members. Wherein employee owned devices drive productivity and flexibility, organization is left choiceless but to embrace the BYOD trend [74]. Agudelo et al. [3], Crossler and Bélanger [60] and Thompson et al. [214] investigated the aspect of ownership and discovered that the more psychologically ownership is manifested, the more likely mobile device users are to safeguard their devices. Employees will demonstrate greater care and accountability for their own devices than the organization's IT assets [127]. This may be due to the fact that they pay for their devices in addition to Internet connectivity and device maintenance. On the negative side, this creates blatant disregard for security policies compliance among employees and also a challenge for organizations to impose them on employees [19, 103].

2.2.3 Mixed Usage

Personal devices are used for both, personal and work purposes due to convenience and cost-effectiveness. It is beneficial for both employers and employees as employees are familiar with their own devices and employers do not need to provide new equipment for employees. Having all the information accessible on a single device using personal devices for work raises concerns about security and privacy. The proliferation of social media and mobile payment applications, among others, makes mobile devices a repository of confidential and sensitive data. Personal, commercial, government, financial, authentication, and connection or service data are found to be safe on mobile devices [213]. This scenario is exacerbated by the fact that employees also store data belonging to third parties, such as images, chats, and messages, making the employee accountable for securing the data of third parties [60]. The poor management of sensitive data may result in a data breach. Loss of confidentiality, availability, and integrity may have severe legal, financial, and reputational consequences for enterprises. Crossler et al. [63] observed that employees who handle sensitive information as part of their employment are more inclined to engage in security behavior. Tu et al. [218] discovered that using mixed usage as a moderator attenuated the overall effect of perceived efficacy on security policy compliance.

2.2.4 Employees Making Security Decisions

Employees are responsible for safeguarding the security of the devices, and PCs are pre-configured with security and privacy options that are easily modifiable, unlike in typical corporations [30]. In the traditional setting, the corporate servers and workstations are reliably protected, under the complete control of the IT Department. In BYOD, it is assumed that owners are responsible for the security of their personal devices [74, 96]. Employees may need to rely on their own capability to protect their devices, but such an approach may open up more vulnerabilities [214]. It is crucial to inform employees about the consequences of using their personal devices at work and steps to take to mitigate risks.

2.2.5 *Visibility*

With some several employees possessing more than one device, various brands, operating systems, settings and as well as data, IT department has many endpoints to manage, even exceeding the number of desktops. This makes it difficult for the IT department to handle all of these devices [218], and it becomes even more challenging when the devices carry personal information. Moreover, the lack of standardization with ubiquitous endpoints increases the complexity and cost of hardware integration, along with the need for IT support. Recent investigation showed misuse by peers in higher education institutions assisting in the authentication of people who were not permitted in the campus network [156]. As a result, network managers failed to identify illegal users and devices on the campus network, leaving them vulnerable to a data breach.

2.2.6 *Function*

Mobile devices, particularly smartphones, are small in size, have superior processing and communication capabilities, and can host a variety of third-party applications [213]. It enables users to access various services through a single device [15]. However, there are some constraints in their processing ability due to their mobility and size [166]. The small size of devices poses challenges in managing mobile security [239]. Furthermore, smartphones lack comprehensive security features [96, 103], rendering traditional countermeasures ineffective. Employees can stay connected by using these ubiquitous gadgets [60, 224] and constantly transmitting data such as user preferences, site clicks, application downloads and usage, audio and video streams, search phrases, and many more. Thompson et al. [214] underlined that many features in mobile devices operate differently, making it difficult for users to adhere to security policies. The combination of different operating environments, touch screen interfaces, portable size and information sharing capabilities accounts for many security vulnerabilities. In mobile devices, the settings for information sharing controls differ since each application has its own settings [60], and some of these privacy settings are hidden or difficult to utilize. Furthermore, some of these devices lack the sophistication to support antivirus, patches, or security configurations for effective protection [103, 218].

2.3 BYOD Benefits

There are various advantages to employees using their personal gadgets in the workplace.

2.3.1 Easy Access to Information

Using personal devices to access corporate emails or corporate data from anywhere offers various advantages. BYOD offers greater flexibility and increase job satisfaction and performance [78, 115] that the modern day employees crave. This may result in improved response times because they can respond to emails while they are away.

2.3.2 Lower Operational Cost

Employers save money by not paying for high-priced devices and data plans [86] and bear routine maintenance expenditures. Employees are responsible for purchasing and upgrading their own devices; hence BYOD initiatives helps organization financially by eliminating hardware procurement costs.

2.3.3 Increased Productivity

Employee productivity is a major driver in the adoption of BYOD. Employees that are mobile and connected are able to work faster and handle work-related tasks. Approximately 36% of the employees were found to have saved two or more hours by using their devices [87] and provide better customer response times. They are also productive on their own devices because they are more familiar with the technology and layout. This removes the barriers to accessing information that complicated or unfamiliar systems bring.

2.3.4 Satisfied Employees

BYOD keeps employees pleased and comfortable by allowing them to utilize gadgets with which they are more accustomed. It also accommodates to younger workers who have grown up with many of these devices before entering the working world [148]. The ease of using a single personal device for both professional and personal purposes eliminate the need for employees to carry and manage several devices at the same time.

2.4 BYOD Security Risk

Along with the benefits, BYOD also comes with inherent security risks. Growing reliance on personal devices in the workplace has become ubiquitous, resulting in new work practices in enterprises. Employees prefer to work on their own devices rather than on organization-owned computers, and expect to be able to work from anywhere at any time. Employees' use of corporate tools and personal applications on their devices inevitably introduces a number of potential risks [112]. The proliferation of IoT and connected devices brought into the modern office by employees has exacerbated the risk [120, 208]. The BYOD presents more risk than traditional desktop models and presents unchartered territories in the organizations' threat landscape [240]. Organizations face greater BYOD risks now that more employees work remotely due to the pandemic [43, 135, 159, 209]. Kearns [121] explained many security concerns that organizations should address to protect data and the existing controls may not be effective against these security risks. Personal devices security is a huge concern and is highly disruptive to organizations as they get exposed to vulnerabilities due to people and human errors [33]. Devices form a target for cybercriminals due to the data that resides inside it [47]. Employees, particularly digital natives, present greater challenges since they focus more on the benefits of BYOD and ignore the risk to themselves and their employers [18, 232].

For this study, the risks are categorized according to the three crucial pillars of security; People, Process and Technology. Presenting BYOD risks according to the three dimensions shows equal importance among them. People who are responsible in implementing security procedures, processes which ensure the right strategies are in place and lastly technology as solutions that support security. The three pillars are interdependent and the failure to address the needs of one, would compromise the structure as a whole. This is because, even with the most advanced and powerful technology, individuals must still be supervised and adhere to policies, or else they will create vulnerabilities. Besides, the ISO 27001 standard advocates the combination of the three pillars.

2.4.1 People/Employees

People or the employees in this context of research are the key components to consider when implementing cyber security measures as insider threat is more complicated [139]. An insider is a current or former employee, contractor, or business partner who has or has had authorized access to an organization's network, system, or data [94, 212] and they have the required knowledge, authority, organization network structure, and security procedures to access all the organizational network information [81].

Studies show that most of the time, information security breaches are caused by trusted insiders. A disgruntled employee or business partner could pose threat to an organization [30, 32, 103, 142, 143, 198, 241]. Some employees can perform malicious actions on data, applications, and intellectual property if they feel they are being treated unjustly, overlooked for an opportunity, driven by greed or looking to gain a competitive advantage with a new employer. Negligent employees are those who are either unaware of threats or pay attention to risks. Insider threats have the ability to do more harm to companies since they are granted legitimate and privileged access to buildings and resources, as well as the and its processes, allowing them to circumvent security procedures [2]. Big organizations with more than 75,000 workers spent around USD 2081 million (around RM8.7 billion) to deal with insider-related incidents [145, 177]. In some cases, 75% of employees find some technical controls to be invasive and ruin the user's experience [32]. Organizations have to find ways to allow measured BYOD usage growth without compromising security.

2.4.1.1 Complacent Behavior

According to research, employees who work for a long time, 5 years or more, constitute a bigger risk to the organization since they exhibit greater confidence at first, gradually become comfortable, and then become less careful with their security behavior compliance [198]. Privileged users are aware of security controls, locations, and organizational information, as well as how to exploit them, making them a threat to the organization [241]. Young employees are classified as high risk since they are the most potential targets of online phishers. A UK government backed research found that 11% of younger adults (18–25 of age) are more likely to fall victim to phishing attacks as compared to 5% of those above 55 years of age [37]. The dangers of phishing are of particular concern during the pandemic, as the number of phishing emails has increased by over 600 percent since the beginning of 2020, as cybercriminals attempt to capitalize on the fear and uncertainty produced by the COVID-19 pandemic [155]. Employees are typically unaware of the risk posed by their conduct because they are unaware of the organization's security policy [142].

2.4.1.2 Sharing Information Online

Study by Belanger and Crossler [30] found that sharing information, especially through mobile devices is a threat to the organizations. It is prevalent particularly when personal devices containing corporate data are connected to insecure public networks that do not require authentication. Insecure information sharing also occurs when employees and contractors fail to activate protective settings on their devices and communicate information over mobile platforms. The explosive growth of social network service (SNS) has significantly changed the way people communicate at home and at work and resulting in disclosing information that they should not [183]. Sharing sensitive information via insecure tools can result in data leaks, ransomware, and malware, among other privacy and security concerns. This is mainly because these devices are always mobile and susceptible to a variety of attacks [30]. Having confidential data in personal devices pose risk of privacy breaches not only from attackers but also from their circle of family and friends who have access to the device [228]. Organizations may face civil liabilities if the public's confidential information is not handled securely [40].

2.4.1.3 Avoiding Security Tasks

Hovav and Putri [103] found that when employees are loaded with security tasks as stipulated in the security policy, they may perceive it as a hindrance to using their personal device as they please. A security procedure or a task might be perceived as additional responsibilities that are troublesome, such as, expecting one to regularly scan or perform software updates. Some employees believe that these procedures can disrupt their work or cause the devices to slow down. There are employees that view certain countermeasures as intrusive to their privacy, hence they are hesitant to implement them [32]. Many employees are too preoccupied with their work-related responsibilities to be concerned with security issues; hence, they do not perceive the value in implementing security protocols [143]. This way of thinking poses difficulties for organization to get their employees to change their habits to improve IT security.

2.4.2 Process

Processes such as policies and security education are key to effective cyber security implementation and crucial in defining how an organization's activities, roles and documentation, besides providing framework for governance.

2.4.2.1 Policy

The primary purpose of guidelines or policies is to improve security behavior of employees by providing relevant direction and value to the employees within organization. Organizations face challenges from the lack of governance of policies and procedures for securing personal devices [234]. It can result in misinformation or uncertainty, resulting in non-compliance behavior and organizations' information assets are susceptible to compromise or theft. The absence of a policy can be attributed to insufficient management engagement and a lack of understanding regarding the significance of having an effective BYOD security strategy in place. This discrepancy could be dangerous for organizations.

Recent research also shown that some organizations do not have a BYOD policy, and if one exists, it only applies to select user groups and devices [32, 143, 226]. Exempting privileged individuals from policy enforcement may cause them to misinterpret the rules. Employees in non-managerial groups, on the other hand, believe that they will not be harmed by any wrongdoing because of their positions [207]. Unfair policy enforcement toward employees, such as exempting privileged positions from enforcement, may create mistrust and indifference among other employees toward compliance. It is essential that all employees, from upper management to the newest, adhere to the policies.

Approximately 75% of respondents stated that there was a lack of mandate from the organization advising employees on the necessity to perform fundamental technical measures such as applying passwords, locking displays, and updating security software and updates and software. Furthermore, studies by Bello et al. [32] and Crossler et al. [63] found that most organizations are hesitant to establish a BYOD security strategy since it is difficult for organizations to enforce controls on employee-owned devices. Bello et al. [32] disclosed that the absence of effective policies, standards, and processes might be a significant obstacle for top management when attempting to manage BYOD. Results also show a worrying reality that even when organizations have policies in place, there are evidence that employees oblivious of its existence [63].

Other issues regarding how policies are drafted in organizations highlighted in the studies are misinterpretation of policy and deliberate engagement in non-compliance behavior [103, 207]. A study conducted focus group discussions with employees of various positions and discovered that employees' interpretations of the policy vary [207]. When it comes to lower-ranking employees, the policy is always interpreted differently based on how it is enforced by their superiors, whereas it is clear for higher-ranking managers. There are also ramifications for organizations whose policies prohibit too many activities that leave employees without control over their devices. According to a study by Hovav and Putri [103], such circumstances will cause employees to disregard security standards and purposefully participate in noncompliance behavior. It can reflect mistrust particularly towards the non-managerial employees. Stringent security regulations may be counterproductive and have an impact on staff compliance levels.

2.4.2.2 Education

BYOD training is crucial in improving employees' privacy security awareness [48], as employees unfamiliar with basic BYOD security best practices can create countless opportunities to compromise organization's information [32]. Bello et al. [32] discovered that 42% of employees lacked knowledge of mitigation techniques and solutions if their devices were attacked or exposed to threats. Employees that are poorly trained in fundamental security measures would be unable to identify risks, detect security breaches, and take corrective action when they occur. Apart from employees' lack of knowledge and skills to handle any form of security breach, the working environment also influence their decision to comply with BYOD security policy requirements [32]. Moreover, mobile device users rely on their own knowledge to safeguard their devices, whereas organizational computing systems are supplied with more robust security solutions [95, 219].

Lack of self-efficacy prohibits employees from enabling BYOD security safeguards that could lead to improper data handling, according to [30]. This is especially important when employees interact with location-based applications and systems, as the amount of personal information acquired and shared unknowingly creates vulnerabilities.

Employees who are ignorant to risk would underestimate the security risk posed by their own device [142]. Unawareness of the security risks emanating from personal devices leads to lax practices in the workplace as employees do not update their devices and software, or use advanced security protections to protect them. Understanding the consequences of threat or perceived threat severity is crucial to improve security policy compliance [63]. This may be especially pertinent for employees who handle sensitive personal data, as security breaches can cost organizations their reputation, financial stability, and customers' trust [63].

On the other hand, Sebescen and Vitak [198] demonstrated contrary findings indicating as technical skills rise, risk reduces. Instead, they discovered that security knowledge and ability were minimal in impacting the highest and lowest risk categories. Overconfident employees, they argued, are more likely to make careless mistakes than less-confident ones.

2.4.3 Technology

Technology, encompassing both hardware and software, is crucial for organizations to achieve robust cybersecurity. While these tools offer significant advantages in protecting against cyber threats, they also bring challenges related to security, governance, processes, and even organizational culture [187]. Despite their potential to enhance security, these technologies can sometimes introduce vulnerabilities, complicate governance, and create operational complexities, all of which require careful management to ensure their effectiveness.

2.4.3.1 Device Management

This topic focuses on issues pertinent to the administration of devices. Today, employees' mobile devices are an integral part of their everyday life and cannot be separated from them [32, 111, 143]. Despite the resistance of some firms to adopt BYOD as part of their everyday operations, employees strongly prefer to utilize their own devices [32]. On the other hand, some organizations permit work-related data to be stored on their employees' personal devices, which could lead to vulnerable situations in which confidential data is compromised [143]. As the devices are small and portable, they can be easily forgotten, misplaced, or stolen (Tu et al. 2015). Sometimes, employees are discouraged from reporting their lost or stolen equipment because it could severely impact the organizations' reputation and financial standing [143]. Lost and stolen mobile computing devices such as laptops and smartphones which contain organizational information can cause information leakage [2]. Some of the devices carrying sensitive data are recycled or sold without ensuring that the data is permanently destroyed.

Additionally, the device's smaller screen size makes it harder for a user to verify the legitimacy of a page or the selected URL links, making it a potential phishing target [111].

The IT teams are only aware of a percentage of the personal devices that employees have opted to use instead of organization-owned equipment. Without knowledge of the number of devices employing a multitude of operating systems and other technical specifications, it is difficult to provide total security against threat. Utilization of unauthorized devices and software produces a shadow IT phenomenon within the organization [142]. Shadow IT refers to all hardware, software, or any other technological solution used by employees without the approval of the IT Department [199]. Finding the correct balance between BYOD and security controls is difficult because extremely stringent measures may negatively influence employee experience and lead to Shadow IT.

Devices are increasingly connected to applications, storage or other digital services, resulting in greater risks [157]. Most devices have Wi-Fi enabled without the user's knowledge, putting them at risk of connecting to rogue networks or public hot spots [241]. Free Wi-Fi or public hotspots do not require authentication to establish a network connection, creating a chance for a hacker to obtain unrestricted access to data from unprotected devices on the same network [111]. A survey done by TrendMicro found that 39 percent of the employees always use public Wi-Fi without Virtual Private Network (VPN), exposing their browsing and passwords to eavesdroppers [101].

2.4.3.2 Application and Data Security

This theme encompasses issues detailing the threats posed to data stored on or accessible by mobile devices [32, 103, 111, 143, 158]. Mobile devices' explosive growth is an opportunity for malware developers to exploit. Furthermore, incorporating security into the design of apps is not the developers' primary priority, since studies have showed that even paid apps on the Android and iOS platforms contain vulnerabilities. This is further exacerbated by the 65% of Android users that have admitted to preferring pirated software over authentic ones [158].

Contrary to organization-owned devices with security apps and controls, personal devices are vulnerable to harmful software and links. These devices are susceptible to malware attacks mostly because their users do not employ mobile antivirus software

[143]. Many employees are naïve or ignorant of the importance of installing antivirus software on their mobile devices. Malware can be easily spread through malicious links in tampered websites, spam emails and messages [111]. These links when clicked would take the victim to a phishing website that steals their critical data or infect their devices. The malicious program is then installed in these unprotected devices, allowing it to easily breach security perimeters and introduce malware into the corporate network. Imgraben et al. [111] discovered that devices employing the 'Remember Me' function to save passwords and credentials are at risk of being compromised because the feature enables attackers to quickly get information from the devices.

Fourteen percent of BYOD users share their passwords with coworkers, family members, and friends due to the fact that employees utilize their devices for both personal and professional purposes. Some employees claimed that, depending on the circumstances, they may share certain information. In addition, respondents in the study did not employ any type of screen locking or password protection as a security measure [32]. This is in line with Imgraben, et al.'s [111] results, that more than 50% of smartphone users lack a locking mechanism. This can have devastating consequences, if the phones are lost, misplaced, or stolen, as all the information saved on them is accessible to others and identity theft is a concern. Mani et al. [143] validated the findings when their inquiry uncovered shocking revelations regarding supervisors who never change their passwords and who share passwords with coworkers so that they can respond to emails in their absence. Research also shows that employing weak passwords are one of the leading vulnerabilities that lead to data breaches and other types of cyber-attacks [88, 165].

Furthermore, mobile devices are not particularly secure, posing a threat to the organization. Consumer-developed technologies lack crucial characteristics, such as manageability or the capacity to conform to security requirements [112]. The detection of malware and spyware on devices is still ineffective. It is challenging to integrate anti-malware and anti-spyware on mobile devices due to their software platform limitations [166]. A device that is infected could potentially compromise the organization's systems and the personal information stored on the device [103].

Employees install illegitimate software in their devices without knowing that they are unauthorized solutions [76]. Unauthorized software widens the attack surface for attackers because it is probably unmanaged and not receiving the necessary patches, updates, and customizations [93]. Illegal or pirated softwares, games, media players, chat applications are some of the examples of unauthorized solutions. With mobile devices, employees install software directly on the device and provide them extensive permissions, allowing hackers to operate with relative ease. Cybercriminals continue to research new ways to exploit vulnerabilities in apps, operating systems, and software, hoping to profit from security weaknesses before manufacturers discover and fix them. Organizations have little control over their employees' device usage, leaving the whole burden to the employees.

The rising popularity of cloud storage, and its ever-increasing versatility has opened more opportunities for employees practicing BYOD. BYOD drives the data to the cloud due to the flexibility, scalability, and the cost-effective of the cloud services. With the cloud services being a regular part of work, organizations must deal with security issues such as loss of control over sensitive data. This is because once the data is outside of the organization's IT environment, the data's privacy settings are beyond the control of the organization.

2.5 Information Security

The safeguarding of information and the systems that use, store, and transmit it is referred to as information security [235]. Organizations need to safeguard information that is of strategic importance, for example personal details, trade secrets or intellectual property, sensitive information and information subject to legal and regulatory requirements [2]. Threats to information systems consist of inadvertent or intentional unauthorized access, disclosure, modification, or destruction of information [136] and they are still relevant today from the last 30 years.

Organizations utilize technological controls to secure the availability, integrity, and confidentiality of data, which are the three most important information security controls. However, Information System (IS) security problems cannot be handled by technical controls alone [167]. Technology controls must be supported by formal controls which refers to the construction of correct security responsibility and authority structures, an all-encompassing strategy and policy, and adequate business procedures. Additionally, it is necessary to determine the roles of the appropriate individuals inside a company in order to preserve accountability, strategy and policy, and business processes. Even formal controls cannot function by themselves [41]. To ensure that formal controls are effective, it is crucial that individuals adopt and accept them.

Existing research has investigated the causes of noncompliance with information security [41, 66, 99, 109], and there are various reasons that decrease employees' compliance. Among those, employees finding security procedures to be troublesome and affecting their productivity [108] especially in a BYOD environment, they are expected to actively share critical information resources, downloading information and applications to complete work task faster.

2.5.1 Information Security Policy

The building blocks of a comprehensive IS security program include security policies, guidelines, standards, baselines and procedures [51]. IS security policy is a component of the formal control structure and it can be identified as a technical or a managerial policy [29]. A policy describes the security duties and responsibilities as well as the standard operating procedures for safeguarding the information resources [41]. It also regulates the human actions regarding information security or expressing the organizations' security aims [171]. Organizations develop security policies to guide and assess employees' behavior in order to prevent negative impact from security breaches [149].

The ISP is issued by the IT department and covers solely IT assets [22]. As a result, an IS security policy is described as a description of the IS security tasks or IS security related activities that employees should perform to ensure that their organizations' information is safeguarded. The right use of computer and network resources, good password practices, often backing up files, checking for encryptions, manual virus check, and locking computer are all part of the IS security tasks [7, 99]. At a technological level, they provide guidelines for implementing systems and processes within the organization and at the user level, these policies provide appropriate behavior attributes for employees.

It is a declaration of the desired level of security and uses terms such as security objectives, strategy, and goals. An ISP provides security guidance to organizations in order to help them implement information security management. A written policy can help safeguard the organization in the event of a legal disagreement regarding penalties.

2.5.2 Information Security Policy (ISP) Compliance

ISP compliance refers to an individual's execution of IS security tasks to preserve IS security in accordance with IS security policies [45, 99]. It points out to behaviors that are required or expected as indicated in security policies [65]. Padayachee [172] defines compliant information security behavior as a set of information security actions that employees must adhere to in order to safeguard information security. Embracing compliance behavior increases the trust and confidence in the capability of the organization besides meeting the security requirements of stakeholders.

Many security issues occur in organizations as a result of non-compliance by personnel. This displays a lack of knowledge or disregard for security policies, as they are unaware of the security implications of their conduct. A study by Albrechtsen [6] indicated that employees may choose not to undertake numerous IS security activities for a variety of reasons, including convenience and inconvenient because people prioritize other responsibilities over security tasks [100]. This is because IS security task demands additional time and effort [41].

Numerous research has been conducted to explain the individual security compliance intention and/or actual behavior with prominent theories by scholars and practitioners; deterrence theory [50, 119, 152, 173, 221], protection motivation theory [61, 132, 192, 222]; theory of planned behavior [41, 99, 110, 152], rational choice theory

[41, 105] and neutralization theory [27, 201]. The use of these theories has enhanced our comprehension since they explain the motive behind individuals' security policy compliance. These studies, however, primarily focused on the traditional computing environment in organizations. Traditional computing is constructed around physical hardware owned by the organization. Starting with a group of desktop computers, these are connected wirelessly or via cable to a network switch and often a server of some sort. As such, any employee working in that office will be able to access the stored information via their computer [123].

2.6 BYOD Security Policy in the Public Sector

With 1.7 million employees, the Malaysian Public Sector is the largest organization in the country [1]. This sector is responsible for administering the Malaysian government (Public Service Department, PSD) (Ahmad and Saad 2020). It strives to supply citizens with goods and services, achieve responsibility, and comply with legal standards, but not to maximize profit. The public sector is responsible for this because there is an efficient and economical exploitation of public fund or resources in accordance with government objectives. The public sector plays a crucial role in determining sustainability and growth of the nation. The use of technology to reduce human interaction and improve service delivery is a key instrument for the government to employ.

This study focused on the Critical National Information Infrastructure (CNII) agencies. CNII is formally addressed in the National Cyber Security Policy (NCSP) which was developed by the National Information Technology Council (NITC) and approved by the Malaysian Cabinet in 2006. To date, NCSP is under the purview of the National Cyber Security Agency (NACSA) [191]. CNII operate numerous physical processes that generates essential telecommunications, power, defense, finance and banking, and more and because of the critical and interdependent nature of CNII, cyber security policies and procedures must be established.

The potential for malicious insider actions to interrupt services and cause physical and reputational harm makes the insider threat one of the most serious security concerns for these organizations [31]. They can pose a major threat to the organizations by revealing or exposing sensitive information [122]. In the government, the majority of people involved in incidents when confidential or proprietary information was stolen are existing employees [146]. This is because most of network security protocols concentrate on limiting access from outside the network. An employee who has been granted authorized internal access has already surpassed the more challenging task of getting external access to a network [146].

The essential industries such as water, defense, energy, and transportation use systems with high security requirements. When these systems are connected to the Internet, cyber criminals can take advantage of them to gain access to their systems. As a result, this must be addressed in order to ensure that the CNII sectors are secure, robust, and self-sufficient. In the Malaysian public sector, government regulatory supports are the directives or circulars from Chief Secretary of the Government (KSN), federal ministries or central agencies such as MAMPU (The Malaysian Administrative Modernisation and Management Planning Unit, Prime Minister's Department), regarding the implementation of specific policies or procedures for adoption, along with a set of specific deadlines [194]. As the public sector's leading agency for ICT planning and development, MAMPU is responsible for ensuring that ICT projects are linked with the country's development strategy. In response to such directions, a series of events will ensue in which agencies urge their internal departments to take the appropriate action. The directives or circulars could be mandatory, optional, or serve as compliance guidance for agencies. For mandatory directives, agencies must adhere to the stated norms and principles. Security policies are one of the mandated directives from MAMPU to all agencies in the public sector.

Security policies are directives, regulations, directives, rules, and procedures that govern how an organization maintains and protects data [205]. The ISPs are crucial to implement security controls and determine employees' adherence to it [82]. In order to be effective, organizations must ensure that their staff comply to these policies [227]. Previous research found that government employees have compliance issues with the ISP due to weaknesses in the enforcement. The research also found that although security policy forbids employees to bring their personal laptops and USB mobile broadbands to use social media at workplace, it still happens among employees [2].

BYOD security policy management in the Malaysian public sector is still in a nascent stage. Although the BYOD practice has been around for around a decade, a specific policy to manage BYOD is still in progress. The ISP contains guidelines that must be read and followed when using Information and Communication Technology (ICT) assets, but it does not cover BYOD policies specifically [140]. Currently, most of the agencies are using the general ISP with few clauses dedicated for BYOD as a reference for all employees. BYOD is a major concern in the public sector as it involves data security with classified official information residing in the mobile devices [44]. Besides, employees have compliance issues with ISP due to weak enforcement [211]. There is a critical need for thorough deployment and implementation of BYOD policy in public sector organizations and institutions [161].

2.7 Related Studies on BYOD Security Policy Compliance Factors

Table 2.1 details the studies on BYOD security policy and Table 2.2 demonstrates prior research that has examined BYOD security.

Table 2.1 List of BYOD security policy studies

Author	Context	Methodology	Theory	Sample	Findings
Belanger and Crossler [30]	Attitude of iPhone users towards sharing information, using protective settings and actual practices	Survey	Theory of Planned Behavior	228 iPhone users	Intention to protect mobile information leads to actual privacy settings practice, and attitudes toward information sharing and self-efficacy in mobile privacy protection influence this intention. Mobile privacy concerns and trust in the mobile platform are determinants of attitude toward information sharing
Chen and Li [48]	Understanding motivation factors in predicting mobile device users' voluntary adoption of security software	Survey	Technology Threat Avoidance Theory	284 users	Privacy concerns and coping appraisal have a substantial effect on the intent to deploy security defensive software

(continued)

Table 2.1 (continued)

Author	Context	Methodology	Theory	Sample	Findings
Crossler and Bélanger [60]	To understand usage of privacy settings on smartphones through iPhone users	Survey	Information–Motivation–Behavioral (IMB) skills model	334 iPhone users	Personal motivation is one of the most significant determinants of privacy-protective settings usage, while social motivation is insignificant. Knowledge (i.e., knowledge specific to the device's privacy settings) and self-efficacy constructs determine one's use of privacy-protective settings
Crossler et al. [63]	Examining the factors that determine whether employees follow BYOD policies	Survey	Protection Motivation Theory	444 students and employees (accounting and non-accounting)	Self-efficacy and response efficacy were the primary drivers of compliance intentions with a BYOD policy. Threat severity was more salient for accountants

(continued)

Table 2.1 (continued)

Author	Context	Methodology	Theory	Sample	Findings
Dang-Pham and Pittayachawan [72]	Examining the intention to perform malware avoidance behaviors among higher education students	Survey	Protection Motivation Theory	252 students	Intention to perform malware avoidance behavior differed across the contexts. Developing community of practice and repeated trainings to maintain the users' confidence in their own abilities to cope with malware threats
Giwah et al. [96]	Investigating information security behavior of mobile device users in the context of data breach	Survey	Protection Motivation Theory	390 mobile users	Practitioners must create information security training programs with a focus on mobile self-efficacy. High impact of MSE on users' protection motivation (PM) to protect their mobile devices
Hovav and Putri [103]	Employees' intention to comply with organizational BYOD security policies	Survey	Reactance, Protection Motivation and Organizational Justice Theories	230 employees	Response efficacy and justice positively impact BYOD policy compliance intentions. The presence of a BYOD support team enhances employee's perceptions of response efficacy and perceived justice

(continued)

Table 2.1 (continued)

Author	Context	Methodology	Theory	Sample	Findings
Thompson et al. [214]	Roles of social influences and psychological ownership, and by including actual behavior of home computer and mobile device users	Survey	Protection Motivation Theory, added with social influences and psychological ownership	629 home computer and mobile device users	Perceived vulnerability, self-efficacy, response cost, descriptive norm and psychological ownership influenced personal computing security intentions and behavior for both home computer users and mobile device user
Verkijika [224]	Understanding smartphone security behaviors in South Africa	Survey	Protection Motivation Theory added with anticipated regret	428 smartphone users	Perceived vulnerability and perceived severity significantly influenced anticipated regret positively. Furthermore, self-efficacy and anticipated regret revealed strong direct connections with smartphone security intents, whereas smartphone security intentions and anticipated regret positively influenced smartphone security practices

(continued)

Table 2.1 (continued)

Author	Context	Methodology	Theory	Sample	Findings
Tu et al. [218]	Extends protection motivation theory by integrating moderating factors related to mixed use and surveillance visibility	Survey	Protection Motivation Theory	122 mobile users	The intention of employees to comply is influenced by their perception of the device's effectiveness for both personal and professional use
Ameen et al. [13]	Understanding employees' information security behavior by considering influences on the national, organizational, technological (smart-phone-specific), and personal levels among employees aged 18–35 from United Kingdom (UK), United States (US)	Survey	Protection Motivation Theory, General Deterrence Theory and Theory of Reasoned Action	300 Gen-Mobile workforces working in international companies	National cybersecurity policies, cultural differences between countries, and risks particular to smartphone use are all focal points for understanding employees' smartphone-security behavior

(continued)

Table 2.1 (continued)

Author	Context	Methodology	Theory	Sample	Findings
Ameen et al. [12]	Understanding gender differences in smartphone security behavioral intention among employees in the United Arab Emirates (UAE) and the United States (US)	Survey	Protection Motivation Theory, General Deterrence Theory and Hofstede's cultural dimensions	1156 employees	There are gender variances, however neither male nor female employees in any of these countries are aware of the hazards associated with their smartphone use, while being aware of their company's BYOD security policy
Das and Khan [74]	Examine the information security behaviors of smartphone users in the Middle East	Survey	Health Belief Model	500 users	Efficacy of security measures and the cost of adopting are main factors influencing smartphone security behaviors. Users are more concern about malware and data leakage than targeted information theft

Table 2.2 Studies relating to BYOD security

Author	Context	Methodology	Theory	Sample	Findings
[10]	Focuses on smartphone security among employees	Expert review	None	None	Smartphone security awareness is influenced by smartphone productivity levels, pressure applied on employees to raise the amount of effort necessary to accomplish work with smartphones, and policy pressure
[32]	Identify and manage bring your own device (BYOD) security and privacy risks faced by organizations that use mobile devices as part of their business strategy	Case study (interviews and survey)	None	62 employees	Identifies BYOD risks, threats, and effects, as well as appropriate controls and processes for managing corporate and personal information resources
[26]	Examine the adaption strategies managers have adopted to deal with the BYOD issues. Coping Model of User Adaptation (CMUA) in the organizational decision-making context is operationalized	Survey	Coping Theory	337 employees (CEOs, top managers or executive and managers in charge of their own service/ department/ business units, with at least one employee to manage, with hierarchical responsibility	Those who have already established steps to control BYOD usage have different coping strategies than those who have not
[20]	A theoretical analysis reversed IT adoption logic vs traditional IT	Theoretical review	None	None	Employees enjoy benefits of using their own devices, but this poses a greater risk to the security of their personal data

(continued)

Table 2.2 (continued)

Author	Context	Methodology	Theory	Sample	Findings
[53]	Investigating intention of adopting BYOD is due to the accessing of security policy by threat and coping appraisal	Survey	Technology Threat Avoidance Theory	450 employees	The main influencers on employees' intentions to adopt BYOD are organizational commitment and job security, while perceived cost and privacy protection have no substantial effect
[111]	Investigate students' security behavior, BYOD risks awareness	Survey	None	250 students	There is a need for continual fundamental online security training programs and the cultivation of a security culture
[142]	An investigation about the influence of Shadow IT usage on organizational knowledge sharing	Interview	None	10 IT users	Shadow IT occurs when organizations do not provide suitable tools to communicate efficiently and it can facilitate knowledge sharing especially when people are geographically distributed
[143]	Investigate BYOD security risks management standards, threats awareness in the real estate sector in South Australia	Interviews and survey	None	20 employees for survey and 20 employees for interviews	Lack of understanding about risks and emphasize the need for training programs

(continued)

Table 2.2 (continued)

Author	Context	Methodology	Theory	Sample	Findings
[158]	Explore the security awareness of smartphone users who download applications from official application repositories (e.g., Google Play, Apple's App Store, etc.)	Survey	None	458	Lack of awareness as majority of users trust the app repository, security controls are not enabled or not added, and users disregard security during application selection and installation
Jarrahi et al. [113]	Outlines changes from the three aspects of the framework (Technology, People and Practice) from two sets of empirical studies	Interviews	None	Study 1 – 24 employees Study 2 – 58 consultants	Three approaches to IT infrastructure management (passive, reactive and pragmatic)
Alaskar and Shen [5]	Focuses on five main contextual factors: device ownership, place, time, activity/task, and sensitivity	Theoretical review	None	None	From a work/life viewpoint, a framework of BYOD contextual variables and the impact of each contextual aspect on employees' compliance with information security policy is described
Sebescen and Vitak [198]	Evaluates how three sets of employee characteristics (demographic, company-specific, and skills-based) predict an employee's likelihood of becoming a security breach victim	Survey	None	250 employees	Individual factors rather than skills have the largest effect on an individual employee's risk level. Being fully aware of their knowledge and skills, they may become overconfident and not take advantage of available learning opportunities; hence causing risk in the organization

(continued)

Table 2.2 (continued)

Author	Context	Methodology	Theory	Sample	Findings
Doargajudhur and Dell [79]	Combines the job demands-resources model and task technology fit model	Survey	Job demands-resources (JD-R)	400 employees	BYOD indirectly affects job satisfaction, job performance, organizational commitment, job demands, job resources and task technology fit
Stephens and Ford [207]	Examines the challenges of having mobile device policy to reduce misuse in organizations	Focus group	None	39 employees	Challenges employees face when interpreting a restrictive mobile device use policy. This tension is manifested through four unintended consequences of the strategically ambiguous policy: (a) declining productivity, (b) supervisors burdened by overwhelming tasks, (c) decreased professional mobility, and (d) fleeting organizational identification
Vrhovec and Markelj [226]	Examines the relation between the use of mobile devices, adhering to hospital information security policy and perceived consequences of data breaches	Survey	None	95 employees	Perceived personal consequences are negatively correlated with personal and work mobile device use for work. Adhering to information security policy is positively correlated with perceived data breach consequences for both the patients and the hospital

2.8 Theories Used in Previous Studies

Literature review found several theories that were used to assess BYOD security policy compliance behavior. Researchers employed integrated theories or a single theory with extra constructs from different models. Previous studies showed theories that explored user behavior towards BYOD security, such as the Protection Motivation Theory (PMT), Theory of Planned Behaviour (TPB), Health Belief Model and Technology Threat Avoidance Theory.

Rosenstock [188] developed the Health Belief Model (HBM), which has been widely applied to health behavior research [36]. According to HBM, people will be more careful and thus practice healthy behavior if they believe in the threat of an illness or disease and know how to prevent them [91]. As depicted in Fig. 2.1, individuals determine the feasibility, benefits, and costs associated with an intervention or behavior change based on perceived susceptibility (similar to perceived vulnerability), perceived barriers, perceived benefits, cues to action, and perceived seriousness (severity), according to HBM.

People will be driven to adopt information security behavior if they perceive a high vulnerability level in information security. It has also been used to predict computer security behavior [163, 222] and to assess the likelihood of a person being exposed to a harmful threat [237]. Perceived severity, on the other hand, relates to an individual's assessment of the significance of a situation [42]. As stated by Rosenstock [188], one can gauge the severity of a certain health issue based on the amount of emotional reactions elicited by the concept of sickness, as well as the kinds of challenges it will present. Likewise, perceived severity has been observed in information security studies to deter employees' violation behavior towards ISPs

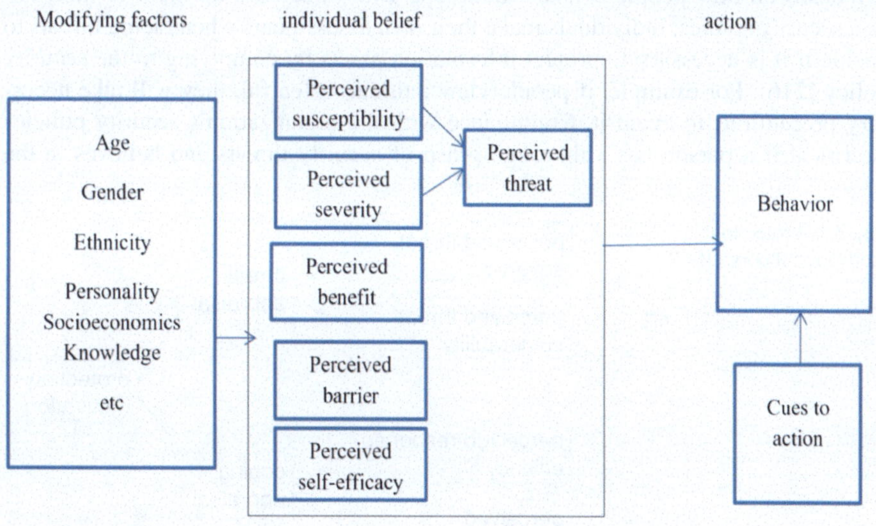

Fig. 2.1 Health belief model [188]

[52]. Huang et al. [106] suggested that modifying users' threat severity perceptions could improve their security compliance behavior.

The **Protection Motivation Theory (PMT)**, which is an adaptation of HBM, was introduced by Rogers in 1975 as a framework to provide clarity to the understanding of fear appeals. This theory describes how people's attitude and behaviors towards health change in response to health risk message. In 1983, the theory was revised to provide a more general perspective of the impact of persuasion communication with an emphasis on its efficacy to mediate behavior change. The PMT concept involves any threat for which there is an effective recommended response that can be carried out by the individual [85].

As indicated in Fig. 2.2, PMT is based on five factors that are considered to motivate users to protect themselves; severity, vulnerability, response cost, self-efficacy and response efficacy. Threat evaluation (threat vulnerability and severity) and coping appraisal (response efficacy and response cost) are two categories of the factors. According to threat appraisal, if a person has a strong perception of the threat severity and vulnerability, it can motivate them to avoid security incidents [130, 182]. Coping evaluation relates to a person's ability to avoid security risk and their belief that they can correctly execute the required security behavior [176, 182].

Information systems security research regularly utilizes PMT to understand an individual's decision to engage in secure behaviors or follow security policies [16, 45, 62, 99, 109, 116, 131, 134]. PMT has been one of the most extensively used theories in the information security research domain due to the variety of threats and associated responses. PMT is one of the most powerful explanatory models used by researchers for explaining human information security behaviors [224]. It is said to be an established, robust theoretical foundation for the analysis and exploration of recommended actions or behaviors to avert the consequences of threats [116]. It focuses on how people consider and cope with threats. In the field of information security studies, individuals make their own assessments when facing threats to decide if it is necessary to protect information assets by complying to the security policy [216]. For example, if people view a threat as fearful, they will take necessary precautions to avoid it. Compliance with the organization's security policies is critical if a person has a thorough grasp of security threats and believes in the

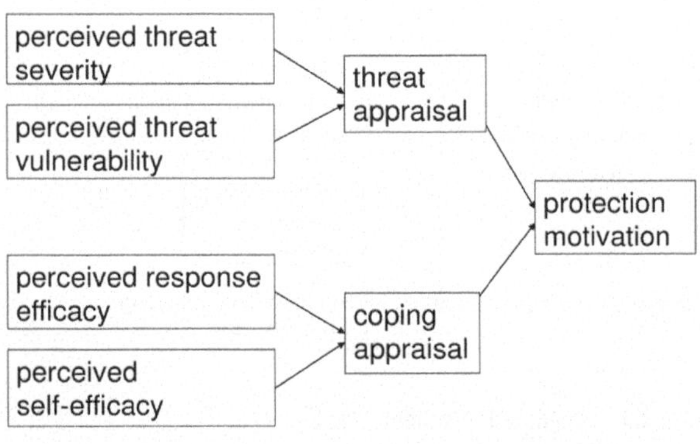

Fig. 2.2 Protection motivation theory [186]

effectiveness of security policies as a coping mechanism [99]. For instance, PMT has been used to explain whether individuals: (1) plan to use anti-malware software [116, 134], (2) intend to adopt anti-plagiarism software [130], and (3) intend to comply with information security policies [99]. PMT variables, namely, perceived severity, self-efficacy, and perceived vulnerability/ susceptibility have a substantial impact on users' security behavior [52, 180, 222]. The literature shows that individuals do not comply blindly with ISP, but make their own assessment and decisions before performing a security policy related secure behavior. In their study on the effect of organizational commitment on insider security behavior, Posey et al. [178] considered threat severity, threat vulnerability, self-efficacy, and response cost. Johnston and Warkentin [116] developed fear appeals model using the PMT components of threat severity, threat susceptibility and self-efficacy.

In the field of BYOD information security policy, Crossler et al. [63] investigated the factors leading to employees' intention to comply and their actual compliance with BYOD policies using the PMT. Their examination included accountants who dealt with sensitive material and non-accountant workers. Using PMT, Dang-Pham and Pittayachawan [72] and Thompson et al. [214] evaluated security perceptions and behaviors between users of mobile devices and home desktop computers. Belanger and Crossler [30], Blythe and Coventry [35], Crossler and Bélanger [60], Giwah et al. [96], Hovav and Putri [103] and Tu et al. [218] used PMT to study BYOD security behaviors among smartphone users. In their investigation of smartphone users' security behavior in South Africa, Verkijika [224] expanded the study on PMT by incorporating the role of anticipated regret. Ameen et al. [12] in their study used PMT to study gender differences in smartphone security behavioral intention among employees in UAE and United States (US).

Theory of Planned Behavior (TPB) is another theory that is proposed by Ajzen [4] used to understand and predict behaviors based on the constraints of theory of reasoned action [4, 118]. As an extension to the Theory of Reasoned Action, TPB was developed with the addition of perceived behavioral control that focuses on the behaviors that that are out of the control of the individual. TPB has been widely used to assess users' adoption of information systems that are intended to anticipate human behavior. Regulations and people's compliance behavior have both been studied using this theory [173, 181].

The TPB places the construct perceived behavioral control within the framework of relationships between belief, attitude, intentions behavior. TPB proposes three conceptually distinct predictors of intention; attitude, subjective norms and perceived behavioral control [4]. Attitude is the extent to which a person has a favorable or unfavorable judgement or appraisal of the behavior. In contrast, subjective norm relates to the perceived social pressure, whereas perceived behavioral control (PBC) is the perceived ease or difficulty of completing the behavior. In general, the more positive the attitude and subjective norm and the greater the perceived behavioral control, the stronger a person's intention to execute appropriate behavior [4]. PBC is thought to influence both intention and behavior as shown in Fig. 2.3. When a scenario or behavior grants a person entire control over his or her behavioral performance, intentions alone should suffice to anticipate behavior.

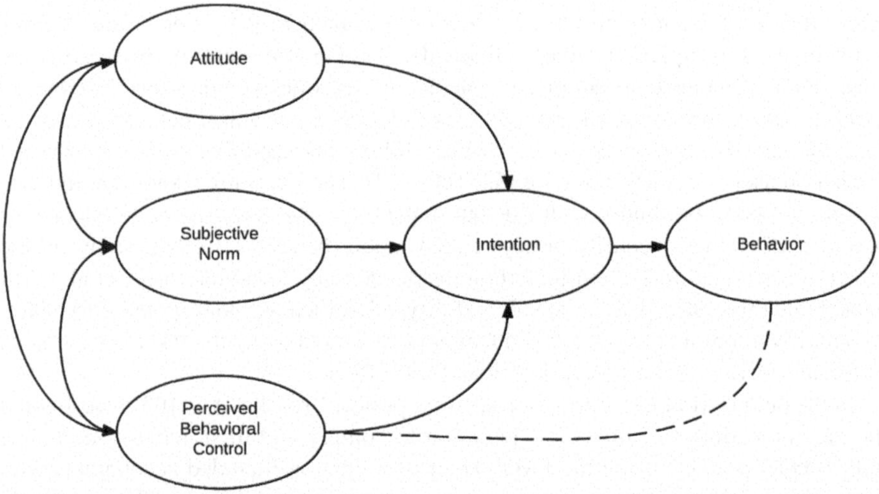

Fig. 2.3 Theory of planned behavior [4]

Several requirements must be met for accurate prediction. As a first step, it is necessary to ensure that the measurements of intention and perceived behavioral control are consistent with or compatible with projected behavior, and he stated context must be same to that in which the behavior of interest would occur. The second prerequisite for accurate behavior prediction is that intents and perceived behavioral control must be consistent between their assessment and the observation of conduct, as intervening events may alter intentions or perceived behavioral control.

Subjective norms are defined by Huang and Chuang [107] as users' perceptions of other people's views on whether or not their actions are appropriate. In this case, other people's opinions are those of superiors and colleagues. However, in this study, the researchers have focused solely on the behavior of superiors due to the fact that employees are impacted by their superiors. Employees tend to abide by what has been taught or show respect to their management, hence the behavior of superiors has the greatest influence on the information security behavior of employees [109].

Meanwhile, perceived behavioral computing can also be defined as users' opinions of their capacity [107], which can be increased by management -provided education and training. According to [210], the PBC construct has two components: Self-efficacy and Facilitating condition. Self-efficacy which derives from Social Cognitive Theory, defines how people feel, think, and what motivates them to behave in a given way based on cognitive, motivational, affective, social influence, and selection processes. Chan et al. [45], state that self-efficacy can be developed by acquiring knowledge. Previous researchers concur that management-provided training programs can boost the self-efficacy of employees. In addition to the correct use of countermeasure tools, self-efficacy also encompasses information security behaviors.

Some scholars admit that additional variables, such as habit, perceived moral obligation, and self-identity, can predict intentions and behavior in the framework of the TRA model, although the TPB did not examine these variables. As an alternative for the TRA's constraint on volitional control, the TPB suggests that behavior is purposeful and planned. Taylor and Todd [210] critiqued TPB because it relies on people being driven to perform a specific behavior, which could pose a problem when examining adoption behavior.

According to Sommestad, et al.'s [203] systematic literature review on the contributing factors on security compliance, this theory was found to be one of the most widely adopted theories by behavioral security researches. TPB factors have been adopted by many behavioral security researches as antecedents of security compliance and other desirable security behaviors [41, 193, 203].

Technology Threat Avoidance Theory (TTAT) formulated by Liang and Xue [134] has been utilized in IS research concerning technology threats. The theory highlights the necessity of understanding users' avoidance behavior when it comes to information technology dangers, and it promotes the idea that users' avoidance behavior is the result of a dynamic positive feedback loop decision process [133]. Users determine how to respond to information security dangers based on the outcomes of threat and coping appraisal. If employees are aware of IT threats, they will avoid risky activity that could jeopardize their careers. According to TTAT, user information technology threat awareness can be influenced by perceived effectiveness, perceived cost, and self-efficacy components [184]. TTAT suggests that to encourage users to avoid malicious behavior, they must recognize the efficacy of security countermeasures and have confidence in their ability to utilize them properly. According to the findings of Mejias [147], knowledge of a malicious IT threat is strongly correlated with information security awareness. Users' awareness of information security would increase if they had a sufficient understanding of IT threats, hence, minimizing the likelihood of information security incidents.

On the other hand, deterrence against deviant behavior is the emphasis of the **General Deterrence Theory (GDT)** and the theory suggests that potential offenders will be discouraged from committing illegal crimes when punishment for behavior is strong and sanctions are harsh. It is based on fear appeal, and when a fear appeal is provided to employees in the form of a policy, they will weigh the benefits and risks of breaching the regulations outlined in the fear appeal [49]. The theory focuses on two primary features of sanctions, namely the severity and certainty of sanctions [149]. In terms of information security behavior, the theory demonstrates that punishment is more effective than norms. Previous studies have combined PMT and GDT to investigate issues in information security research [99, 200]. In one of the BYOD security studies, Ameen et al. [12] deployed both PMT and GDT to provide more informative view of appraisal and fear combined.

The **Psychological Contract Theory (PCT)** posits that employees form expectations and beliefs about their work. A psychological contract (PC) is defined as a person's thoughts on an exchange with another party [17]. In an organizational context, it can be seen as reciprocal obligations and commitments between the employees and the organization. Frequently, promises and intentions are conveyed through corporate policies and practices [189]. The terms of an employee's PC are frequently derived from information provided by the organization. A PC violation occurs when an employee believes that the organization has failed to meet its commitments [57]. As a result, disruptive or rebellious behavior is used by the aggrieved party, when one party feels there is an imbalance in the reciprocal connections [238]. ISPs may increase user obligations without specific reciprocity from the organization. ISPs can therefore be considered a violation of PC. In a BYOD environment, employees may consider the firm's deployment of BYOD ISPs that affect users' personal devices as a new obligation in the absence of additional reciprocal commitment from the organization (i.e., a breach of their psychological contract with the organization). Guest [97] for example, contended that PC cannot be assessed and may exist only in the minds of one party but not the other.

According to Workman [238], users' views of PC breach are influenced by response efficacy, threat perceptions, perceived justice, and reactance. The combination of threat perceptions and organizational justice, according to Workman [238], encourages employees' acceptance of corporate monitoring, even when consumers view surveillance as a breach of PC. When employees feel that a threat claimed by the organization is real (i.e., of high threat severity), they are more inclined to accept the necessity to implement risk-reduction measures by the organizations and will not regard such steps as a violation of PC [238].

Reactance Theory (RT) assumes that individuals have a collection of executable free behaviors. Any termination or threat of elimination of these free actions will elicit a psychological response. Psychological reactance is a state of motivation geared toward the restoration of threatened or lost behavioral autonomy [38]. RT is linked to the notion of perceived control [39] which is defined as an individual's perception of having the ability to affect the probability of a potential outcome to occur. A constraint or deprivation of an individual's control means a threat to or reduction of one's freedom. Freedom is defined by actions that are executable, plausible, or possible [38]. Reactance presents itself as an attempt to reclaim lost or threatened freedom. Freedom can be restored either explicitly or indirectly [38, 39]. Regaining one's freedom is accomplished by engaging in the threatened or discontinued behavior. By engaging in other behaviors that belong to the same category as the threatened or terminated free behavior, freedom is implicitly restored. While the majority of studies have focused on the effect of reactance on individuals' attitude and conduct, only a few studies have studied its causes [77, 83]. In an organizational environment, the implementation of organizational power that substantially restricts an individual's autonomy may result in workplace deviance and the violation of organizational standards.

Table 2.3 shows that Protection Motivation Theory (PMT) is the most frequently used theory (7 studies), followed by the Theory of Planned Behavior (1), Reactance theory (1), Health Belief Model (1) and Technology Threat Avoidance Theory (1). Table 2.4 describes the factors analyzed in extant studies.

Table 2.3 Number of occurrences of behavioral theories used in selected studies

Theories	References	N
Protection Motivation Theory (PMT)	[63, 72, 96, 103, 214, 218, 224]	7
Theory of Planned Behavior (TPB)	[30]	1
Reactance theory	[103]	1
Health belief model (HBM)	[74]	1
Technology threat avoidance theory (TTAT)	[48]	1

2.9 Overview of Underpinning Theories

2.9.1 Organizational Control Theory

The definition of control is "a process, system or environment designed to provide motivation for organizational members to take actions and make decisions consistent with organizational objectives" [75, 114, 125, 170]. Individuals within organization need to be motivated to maintain good security practices at work according to organizational objectives.

Control theory [84, 168] focuses on individuals compliance within the work environment due to management policies and procedures. As reported by Ouchi [168], behavior and outputs resulting from behavior are the only two phenomena that can be observed, monitored and counted when studying control systems. When organizations grow in size, they increase in differentiation that creates a problem of integrating subunits. The control problem is fixed through the application of more managerial manpower [168].

Formal control in organizations is based on a performance evaluation strategy that specifies and evaluates outcomes or behaviors. [84, 124, 169]. The control system consists of the techniques organizations use to monitor and evaluate the behavioral performance of individuals against some standard and to recommend a measuring scheme where either the behaviors or the outputs resulting from human behavior are measured [168, 169]. Reward or punishment is the final component, in which individuals are rewarded or punished for participating in a prescribed action [125].

2.9.2 Security Culture

Information security culture is an aspect of an organization's culture and can be considered as a subculture that emphasizes information security in employees' daily activities [9, 162]. Security culture is a critical component of an organization [70] and relates to the way things are done in the organization to protect information [79] and these can be seen in the organizations' collective values, norms and knowledge and their basic assumption [195, 197, 220]. Security culture is strongly entrenched within

Table 2.4 List of factors analyzed in selected studies

Factor	Similar/equal	Number of occurrences	References
Threat severity	Perceived severity/ perceived data breach consequences/ perceived personal data breach	10	[72, 74, 96, 103, 214, 218, 224, 226]
Self-efficacy	Privacy self-efficacy/ technology self-efficacy	10	[30, 48, 60, 63, 73, 214, 218, 224]
Response cost	Perceived cost	9	[48, 63, 72, 74, 103, 214, 218, 224]
Actual behavior	Adhering to information security policy/ security behavior/ mobile information protection behavior	9	[30, 32, 48, 60, 74, 214, 224, 226]
Response efficacy		8	[48, 63, 72, 74, 103, 214, 218, 224]
Threat vulnerability	Threat susceptibility/ perceived vulnerability	7	[63, 72, 74, 96, 214, 218, 224]
Intention to comply	Assurance motivation/ intention to perform malware/ mobile information protection intention	7	[30, 63, 72, 103, 214, 218]
Awareness	Mobile setting protection awareness/privacy risk awareness/ BYOD security	4	[30, 48, 60, 103]
Privacy concern		2	[30, 48]
Attitude towards information sharing	Sharing preferences	2	[30, 60]
Trust		2	[30, 74]
Prior invasion experience	Prior experience	2	[30, 214]
Subjective norm		2	[60, 214]
Anticipated regret		2	[48, 224]
Descriptive norm		1	[214]

(continued)

Table 2.4 (continued)

Factor	Similar/equal	Number of occurrences	References
User sophistication		1	[74]
Rewards		1	[72]
Psychological ownership		1	[214]
Perceived freedom threat		1	[103]
BYOD IT support		1	[103]
Perceived digital mutualism justice		1	[103]
Knowledge	Privacy knowledge/ technology knowledge	1	[60]
Technical security and privacy controls		1	[32]
Information security standards		1	[32]
Liability in BYOD		1	[32]
Information privacy principles		1	[32]
Awareness and training programs		1	[32]

its organizational culture [71] and it was adapted from Schein [195] by da Veiga and Eloff [68] and Van Niekerk and Von Solms [220]. It is present in an organization both as the dominant culture and as subcultures, with distinct information cultures for each department or job level [67, 70] and encompasses employees' thoughts, emotions, and daily activities.

The values and belief that all levels of an organization's members hold in common regarding information security reflects the organization's security culture [46]. In the context of information security, people are frequently viewed as an impediment rather than an asset, making security culture crucial for shaping employee conduct [90]. Information security culture is a guide and structure to human behavior when interacting with ICT to avoid actions that may cause security risk to the information assets [8]. Additionally, it can also be defined as the awareness and comprehension of security challenges and procedures [175]. Creating a security-conscious culture may lower the danger of malicious employee conduct in enterprises [162].

Creating a culture of information security in a company has been shown by a number of studies to manage and reduce information asset security threats [68, 90, 126, 190, 242]. Security culture is crucial as it can improve security related behavior when employees pose risk to the protection of information assets of the organization [11]. Management is expected to be the main shaper and builder of the organization [104] and their commitment and strong leadership is required in order to succeed in the long term.

It is vital to examine security culture because various workplace practices and cultures, as well as regulatory environments, create a unique challenge in the Pacific Asia region [86]. Although, there is ample theoretical support for security culture focusing on conceptual and frameworks [9, 69, 196], very few empirical studies have been conducted [236]. There is also no widely accepted dimensions for security culture and it is still an evolving area [160].

2.9.3 Social Cognitive Theory

Social Cognitive Theory (SCT) evolved from Social Learning Theory (SLT), which was introduced with the publication of Social Learning Theory by Miller and Dollard. Many researchers contributed to the development of SLT, mainly Albert Bandura. His theory was the first to incorporate the notion of modelling as a social learning form, which was renamed as Social Cognitive Theory (see Fig. 2.4). SCT comprises a vast array of factors that function as regulators and motivators of established cognitive, social, and behavioral abilities. Among the most important factors are Reciprocal Determinism, which means that the human behavior results from the triadic, dynamic, and reciprocal interplay of the environment, personal variables, and behavior. SCT is frequently referred to as a bridge between behavioral and cognitive learning theories, as it focuses on the interaction of internal elements such as thinking and symbolic processing (e.g., attention, memory, motivation) and external determinants (e.g., rewards and punishments) in determining behavior [151]. The theory asserts that individual behavior, subjective cognition, and social environment are dynamic and interactive. Subjective cognition involves two basic elements: self-efficacy and outcome expectancy.

Fig. 2.4 Components of social cognitive theory

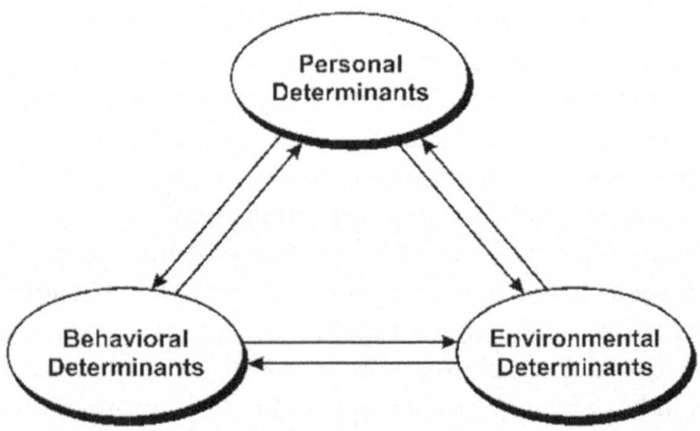

This theory holds that personal and environmental factors work as interacting determinants that have bidirectional effects on human behavior [24]. It also examines how beliefs of self-efficacy influence motivation and behavior [25]. Self-efficacy strength refers to a person's conviction in his or her own estimation of efficacy and resistance to disconfirmation. It refers to a person's awareness of his or her ability to act in the process of reaching a particular objective and a degree of self-confidence in that capacity [138]. Expected outcome is a person's evaluation of the potential outcome of his or her own job. It is a widely established theory that has been used for validating human behavior in a variety of fields, including cancer, IS team development knowledge contribution, and programmer perception of knowledge sharing. Studies have revealed that social cognition has a significant impact on the performance of some human behaviors, especially in the contribution and dissemination of knowledge. Information security research found social cognition also plays a significant role in determining security policy compliance behavior [110, 117, 217]. Bandura [23] claims that low self-efficacy increases the likelihood of people avoiding tasks that are thought to be unfamiliar or difficult to perform. An empirical study by Compeau and Higgins [56] was carried out to develop and validate a measure of computer self-efficacy and to assess both its impact and antecedents. Compeau et al. [55] are recognized as pioneers in IS research for integrating the SCT for the first time to investigate the influence of computer self-efficacy, outcome expectations, and anxiety on computer use. Self-efficacy has been shown to have a significant impact on how people interact with technology. The research findings confirmed the social cognitive impact on computing behavior and demonstrated that self-efficacy had a significant role in shaping the emotions and behaviors of individuals. Individuals with strong self-efficacy utilized computers more frequently, gained greater enjoyment from them, and experienced less worry when using them. The inclusion of technical support was motivated by the author's desire to investigate computer usage in an organizational context. Individuals with strong self-efficacy utilized computers more frequently, gained greater enjoyment from them, and experienced less worry when using them.

2.10 Limitations of Theories Used in Previous Studies and Justification of Selected Theories

Section 2.8 discussed theories used to examine BYOD security policy compliance in previous studies. However, there are limitations associated with these theories. Although, PMT is frequently used for examining user security policy compliance [58], it focuses on fear driven behaviors, leaving out environmental and cognitive factors [13]. As PMT is considered a revised version and expansion of HBM, the usage is quite similar, which does not address environmental factors that can influence one's behavior. Both PMT and HBM relates to a person's perception regarding maladaptive behavior and perceived severity on how bad the threat would be. On the other hand, TPB does not take into consideration emotional factors [14] and change of cognitions unlike SCT with limited predictive ability [202]. TPB models requires one to be motivated to perform certain behavior and this might be problematic when studying compliance behavior.

This study is driven by OCT which explains the processes managers use to steer employees to behave in accordance to the organizational objectives [170] and security culture indicates the values and beliefs of information security shared by everyone regardless of their position in the organization [64]. The study is informed by the literature on SCT [215] which focuses on environmental and cognitive factors in information security that affect employee behavior. Cram et al. [58] recommend that theories like SCT will provide basis for self-efficacy and SETA in the security policy compliance, which is more suitable for security policy compliance study.

OCT is apt to study the public sector information security due to the mandatory requirements of security policy compliance which is a compulsory requirement in measuring public sector organizations' performance. SCT is suitable to study individual behavior in the realm of information security because self-regulated behavior seems critically important for ensuring information security. In this study, we are particularly interested in exploring the degree of influence self-efficacy and BYOD IT support has on BYOD.

2.11 Gaps in the Literature

Seven types of literature gaps were discovered by Müller-Bloch and Kranz [153], Robinson, et al. [185], and Miles [150]. The seven research gaps are as follows:

- Evidence gaps that emerge when new research contradicts earlier findings, necessitating the collection of additional evidence in order to reach a concise conclusion.
- Knowledge gaps is lack of knowledge (e.g., theories, techniques) in each field or the delivery of some unexpected outcomes from studies.
- Practical—knowledge gaps indicate the need for fresh research when there is a disparity between real professional practices and research findings on a particular topic.
- Methodological gaps examine potential conflicts between research methods, the effects of research methods on research outcomes, and the absence of research methods for a certain study field.
- Empirical gaps occur when a specific study area or issue has not been empirically addressed in earlier research.
- Theoretical gaps examine the conflict that may occur when a specific topic is investigated with a single theory or when one theory becomes superior to others.
- Population gaps occur when a subset of the population, depending on race, ethnicity, socioeconomic position, etc., is underrepresented in research.

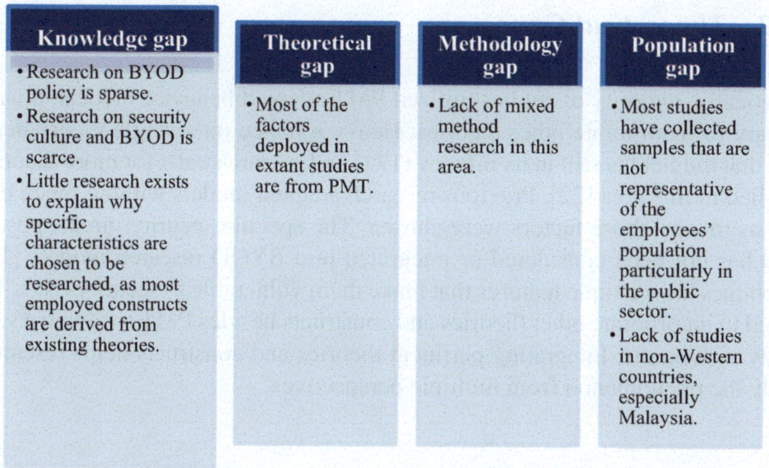

Fig. 2.5 Summary of gaps

Following a review of the literature on information security policy compliance behavior, major gaps were identified, as illustrated in Fig. 2.5. This research attempts to address knowledge, theoretical, methodological and population gap. Each of the gap is explained below.

2.11.1 Knowledge Gap

Existing literature has overlooked elements related to BYOD policies that may have significant impact on how employees' utilization of personal devices affects organizations, resulting in a knowledge gap. Besides, there are only a handful of studies that investigated security policies in a BYOD context [229]. Changing technology and organizational environment requires an update of constructs relevant to the context [54]. In addition, there is a dearth of empirical evidence on security culture [34, 236], and only a few studies have examined the influence of BYOD SETA on BYOD users so far [133]. Employees' BYOD security compliance behavior is a challenge, according to the majority of studies, and the question of additional factors impacting employees' security compliance behavior remains an open one [80]. Extant studies found that organizations cannot rely on technical measures alone to protect their information assets [89, 204, 229, 236] which also emphasizes the necessity to include human security behavior in studies [21, 179, 180]. There is also lack of studies exploring determinants of security policy compliance using control measures that mitigate risk as predictors of compliance behavior.

2.11.2 Theoretical Gap

Theoretically, majority of studies utilized PMT-adapted behavior models; thus, it is necessary to incorporate other theories. Heavy reliance on a single theory demonstrates that the field is still in its infancy [174] and research calls for more theories to be applied in this area [22]. Previous research adapted models without much explanation as to why those factors were chosen. The specific security threats posed by BYOD has not been considered or integrated into BYOD research models [13] as smartphones have unique features that make them vulnerable to security risks. There is a need to incorporate other theories and constructs besides PMT to obtain a variety of study viewpoints. Integrating pertinent theories and constructs helps researchers to study the phenomenon from multiple perspectives.

2.11.3 Methodological Gap

In terms of methodology, most of the literature examined to far relied mostly on diverse statistical analyses to validate the proposed conceptual framework and corroborate the hypothesized causal linkages of the framework's variables, to the best knowledge of the author, none of these studies leveraged the advantages of qualitative and quantitative analyses. The majority of BYOD security policy compliance research has been conducted quantitatively. Almost no qualitative research has been undertaken to investigate compliance factors related to the mitigation of BYOD security risks. Numerous research studies that effectively used qualitative and quantitative methodologies to validate new or integrated theories and test hypotheses have extolled the benefits of mixed method research approaches [59, 223]. This method delivers useful insights from the population, as opposed to those defined by previous study, and enables the emergence of additional factors that may not have been identified by previous studies.

2.11.4 Population Gap

There is a significant gap in the population of the research area. The smartphone penetration among Malaysian was almost 97% in 2021 [154] and smartphones are the most popular devices to access Internet at 98.7% in 2020 [141]. The country has among the highest rate of internet penetration rates in Southeast Asia at 27.34 million internet users. It is important to note that the vast majority of investigations were conducted in Western nations leaving out the Asia region that has received little attention. Although security behavior is widely understood, there is some skepticism about the effectiveness of compliance approaches developed in the Western world when applied to non-western cultures. Most studies were conducted in Western countries and nothing much is known on the BYOD security policy in Malaysia particularly in the public sector. Having different security requirements, regulations, and ICT capabilities from those in Western countries makes the local IT security landscape distinct and important. The current study is among the pioneering studies to investigate employees' BYOD security policy compliance factors in Malaysia. Investigating the organizational context, particularly the government sector, is essential because it may provide contradictory findings about employees' BYOD practices who are responsible for dealing with sensitive, compartmentalized, or classified information. The sector's multidimensional and dynamic environment necessitates the identification and evaluation of risks, as any breached vulnerability could have a crippling effect on national security, national economic security, and public health. Therefore, it is essential to investigate security behavior in 'non-western' countries. Moreover, as a result of the pandemic, the majority of employees are working remotely, which necessitates special consideration owing to their increased cyber risk exposure [159]. As such, this study intends to produce novel findings that reflect Malaysian perspective.

2.12 Summary

This chapter introduces BYOD practices in organizations by discussing the characteristics, benefits, and security threats of BYOD. The chapter reviewed key theories and previous empirical works, enabling the formulation of conceptual framework to be used in answering the research questions. Based on thorough literature review on BYOD security policy compliance studies, several gaps were identified. Existing research employed behavioral theory components or the entire theory to explain compliance intention. However, by identifying risks and elements that minimize risk, it is possible to have a greater understanding of how positive behavior can be encouraged and regulated in the workplace. Most research included random respondents, leaving out the workplace setting, which is more complex considering the organizational and working environment factors. There are very few empirical research exploring links between security culture and security policy compliance. Literature review presented in this chapter leads to the nomination of theories and constructs which is explained in the following chapter. Comprehensive systematic reviews presented in this chapter lead to the identification of independent variables of specification of policy, evaluation of policy, perceived mandatoriness, BYOD IT support, self-efficacy, SETA, psychological ownership, and security culture; and BYOD security policy compliance intention as the dependent variable. These variables extracted from extant theories and studies provide the essential foundation for the development of research framework for this study. Detailed explanation of the framework and development of hypotheses for this research is presented in the next chapter.

References

1. Abas, A.: 1.71 million civil servants on govt payroll as of March 2019. New Straits Times (2019). https://www.nst.com.my/news/government-public-policy/2019/04/484308/171-million-civil-servants-govt-payroll-march-2019
2. Abdul Molok, N.N.: Mitigating the Risk of Organisational Information Leakage Through Online Social Networking. The University of Melbourne (2013)
3. Agudelo, C.A., Bosua, R., Ahmad, A., Maynard, S.B.: Understanding knowledge leakage & BYOD (Bring Your Own Device): a mobile worker perspective. In: 26th Australasian Conference on Information Systems (ACIS 2015), pp. 1–13 (2016)
4. Ajzen, I.: The theory of planned behavior. Organ. Behav. Hum. Decis. Process. **50**, 179–211 (1991). https://doi.org/10.1016/0749-5978(91)90020-T
5. Alaskar, M., Shen, K.N.: Understanding Bring Your Own Device (BYOD) and employee information security behaviors from a work-life domain perspective. AMCIS 2016 Proc. **38**, 1–10 (2016)
6. Albrechtsen, E.: A qualitative study of users' view on information security. Comput. Secur. **26**(4), 276–289 (2007). https://doi.org/10.1016/j.cose.2006.11.004
7. Albrechtsen, E., Hovden, J.: Improving information security awareness and behaviour through dialogue, participation and collective reflection. An intervention study. Comput. Secur. **29**(4), 432–445 (2010). https://doi.org/10.1016/j.cose.2009.12.005
8. Alhogail, A.: Design and validation of information security culture framework. Comput. Hum. Behav. **49**, 567–575 (2015). https://doi.org/10.1016/j.chb.2015.03.054
9. Alhogail, A., Mirza, A.: A proposal of an organizational information security culture framework. In: Proceedings of 2014 International Conference on Information, Communication Technology and System, ICTS 2014, Jan, pp. 243–249 (2014). https://doi.org/10.1109/ICTS.2014.7010591
10. Allam, S., Flowerday, S.V., Flowerday, E.: Smartphone information security awareness: a victim of operational pressures. Comput. Secur. **42**, 56–65 (2014). https://doi.org/10.1016/j.cose.2014.01.005
11. Alnatheer, M.A.: Information security culture critical success factors. In: Proceedings—12th International Conference on Information Technology: New Generations, ITNG 2015, pp. 731–735 (2015). https://doi.org/10.1109/ITNG.2015.124
12. Ameen, N., Tarhini, A., Hussain Shah, M., Madichie, N.O.: Employees' behavioural intention to smartphone security: a gender-based, cross-national study. Comput. Hum. Behav. **104**(106184) (2020). https://doi.org/10.1016/j.chb.2019.106184
13. Ameen, N., Tarhini, A., Shah, M.H., Madichie, N., Paul, J., Choudrie, J.: Keeping customers' data secure: a cross-cultural study of cybersecurity compliance among the Gen-Mobile workforce. Comput. Hum. Behav. **114** (2021). https://doi.org/10.1016/j.chb.2020.106531
14. Ameen, N., Tarhini, A., Shah, M.H., Nusair, K.: A cross cultural study of gender differences in omnichannel retailing contexts. J. Retail. Consum. Serv. **58**, 102265 (2021). https://doi.org/10.1016/j.jretconser.2020.102265
15. Ameen, N., Willis, R.: A generalized model for smartphone adoption and use in an Arab context: a cross-country comparison. Inf. Syst. Manag. **35**(3), 254–274 (2018). https://doi.org/10.1080/10580530.2018.1477300
16. Anderson, C.L., Agarwal, R.: Practicing safe computing: a multimethod empirical examination of home computer user security behavioral intentions. MIS Q. **34**(3), 613–643 (2010)
17. Argyris, C.: Understanding organizational behavior (1960)
18. Baillette, P., Barlette, Y.: Coping strategies and paradoxes related to BYOD information security threats in France. J. Glob. Inf. Manag. **28**(2), 1–28 (2020). https://doi.org/10.4018/JGIM.2020040101
19. Baillette, P., Barlette, Y.: BYOD-related innovations and organizational change for entrepreneurs and their employees in SMEs: the identification of a twofold security paradox. J. Organ. Chang. Manag. **31**(4), 839–851 (2018). https://doi.org/10.1108/JOCM-03-2017-0044

20. Baillette, P., Barlette, Y., Leclercq-Vandelannoitte, A.: Bring your own device in organizations: extending the reversed IT adoption logic to security paradoxes for CEOs and end users. Int. J. Inf. Manag. **43**, 76–84 (2018). https://doi.org/10.1016/j.ijinfomgt.2018.07.007
21. Balapour, A., Nikkhah, H.R., Sabherwal, R.: Mobile application security: role of perceived privacy as the predictor of security perceptions. Int. J. Inf. Manag. **52**, 102063 (2020). https://doi.org/10.1016/j.ijinfomgt.2019.102063
22. Balozian, P., Leidner, D.: Review of IS security policy compliance. ACM SIGMIS Database: DATABASE Adv. Inf. Syst. **48**(3), 11–43 (2017). https://doi.org/10.1145/3130515.3130518
23. Bandura, A.: Perceived self-efficacy in cognitive development and functioning. Educ. Psychol. **28**(2), 117–148 (1993). https://doi.org/10.1159/000180583
24. Bandura, A.: Social cognitive theory. In: Vasta, R. (ed.) Annals of Child Development. vol. 6, pp. 1–60. JAI Press Greenwich, CT (1989). https://doi.org/10.1111/1467-839X.00024
25. Bandura, A.: Social Foundations of Thought and Action. Prentice Hall (1986)
26. Barlette, Y., Jaouen, A., Baillette, P.: Bring Your Own Device (BYOD) as reversed IT adoption: Insights into managers' coping strategies. Int. J. Inf. Manag. **56**(102212) (2021). https://doi.org/10.1016/j.ijinfomgt.2020.102212
27. Barlow, J.B., Warkentin, M., Ormond, D., Dennis, A.R.: Don't make excuses! Discouraging neutralization to reduce IT policy violation. Comput. Secur. **39**, 145–159 (2013). https://doi.org/10.1016/j.cose.2013.05.006
28. Baskerville, R.: Information security: going digital (invited lecture). In: Proceedings of 47th Annual Conference of the South African Computer Lecturers' Association SACLA 2018, vol. 963, pp. 3–14 (2018). https://doi.org/10.1007/978-3-030-05813-5_1
29. Baskerville, R., Siponen, M.: An information security meta-policy for emergent organizations. Logist. Inf. Manag. **15**(5/6), 337–346 (2002). https://doi.org/10.1108/09576050210447019
30. Belanger, F., Crossler, R.E.: Dealing with digital traces: understanding protective behaviors on mobile devices. J. Strat. Inf. Syst. **28**(1), 34–49 (2018). https://doi.org/10.1016/j.jsis.2018.11.002
31. Bell, A.J.C., Rogers, M.B., Pearce, J.M.: The insider threat: behavioral indicators and factors influencing likelihood of intervention. Int. J. Crit. Infrastruct. Prot. **24**, 166–176 (2019). https://doi.org/10.1016/j.ijcip.2018.12.001
32. Bello, A.G., Murray, D., Armarego, J.: A systematic approach to investigating how information security and privacy can be achieved in BYOD environments. Inf. Comput. Secur. **25**(4), 475–492 (2017). https://doi.org/10.1108/ICS-03-2016-0025
33. Bitton, R., Finkelshtein, A., Sidi, L., Puzis, R., Rokach, L., Shabtai, A.: Taxonomy of mobile users' security awareness. Comput. Secur. **73**, 266–293 (2018). https://doi.org/10.1016/j.cose.2017.10.015
34. Blythe, J.M.: Information Security in the Workplace: A Mixed-Methods Approach to Understanding and Improving Security Behaviours, p. 291 (2015). https://core.ac.uk/download/pdf/80693809.pdf
35. Blythe, J.M., Coventry, L.: Costly but effective: comparing the factors that influence employee anti-malware behaviours. Comput. Hum. Behav. **87**, 87–97 (2018). https://doi.org/10.1016/j.chb.2018.05.023
36. Bonar, E.E., Rosenberg, H.: Using the health belief model to predict injecting drug users' intentions to employ harm reduction strategies. Addict. Behav. **36**(11), 1038–1044 (2011). https://doi.org/10.1016/j.addbeh.2011.06.010
37. Brecht, D.: Phishing attacks targeting young adults. INFOSEC (2017). https://resources.infosecinstitute.com/topic/phishing-attacks-targeting-young-adults/
38. Brehm, J.W.: A Theory of Psychological Reactance. Academic (1966)
39. Brehm, S.S., Brehm, J.W.: Psychological Reactance: A Theory of Freedom and Control. Academic Press (1981)
40. Bromwich, M.: Privacy risks when using mobile devices in health care. Can. Med. Assoc. J. **188**(12), 855–856 (2016). https://doi.org/10.1503/cmaj.160026/-/DC1
41. Bulgurcu, B., Cavusoglu, H., Benbasat, I.: Information security policy compliance: an empirical study of rationality-based beliefs and information security awareness. MIS Q. **34**(3), 523–548 (2010). https://doi.org/10.1093/bja/aeq366

42. Bylund, C.L., Galvin, K.M., Dunet, D.O., Reyes, M.: Using the Extended Health Belief Model to understand siblings' perceptions of risk for hereditary hemochromatosis. Patient Educ. Couns. **82**(1), 36–41 (2011). https://doi.org/10.1016/j.pec.2010.03.009
43. Castellanos, S.: With Remote Work Here to Stay, IT Executives Reassess Tech Priorities (2020). https://www.wsj.com/articles/with-remote-work-here-to-stay-it-executives-reassess-tech-priorities-11589825596
44. CGSO Malaysia: Cabaran BYOD dalam persekitaran kerja (2018). https://documen.site/download/kawalan-keselamatan-dokumen-maklumat-yang-di-proses-di_pdf
45. Chan, M., Woon, I., Kankanhalli, A.: Perceptions of information security in the workplace: linking information security climate to compliant behavior. J. Inf. Privacy Secur. **1**(3), 18–41 (2005). https://doi.org/10.1080/15536548.2005.10855772
46. Chang, S.E., Lin, C.S.: Exploring organizational culture for information security management. Ind. Manag. Data Syst. **107**(3), 438–458 (2007). https://doi.org/10.1108/02635570710734316
47. Chatterjee, S., Kar, A.K., Dwivedi, Y.K., Kizgin, H.: Prevention of cybercrimes in smart cities of India: from a citizen's perspective. Inf. Technol. People **32**(5), 1153–1183 (2019). https://doi.org/10.1108/ITP-05-2018-0251
48. Chen, H., Li, W.: Mobile device users' privacy security assurance behaviour: a technology threat avoidance perspective. Inf. Comput. Secur. **25**(3), 330–344 (2017). https://doi.org/10.1108/ICS-04-2016-0027
49. Chen, X., Wu, D., Chen, L., Teng, J.K.L.: Sanction severity and employees' information security policy compliance: investigating mediating, moderating, and control variables. Inf. Manag. **55**(8), 1049–1060 (2018). https://doi.org/10.1016/j.im.2018.05.011
50. Chen, Y., Ramamurthy, K., Wen, K.-W.: Organizations' information security policy compliance: stick or carrot approach? J. Manag. Inf. Syst. **29**(3), 37–41 (2012). https://doi.org/10.2753/MIS0742-1222290305
51. Chen, Y., Ramamurthy, K., Wen, K.W.: Impacts of comprehensive information security programs on information security culture. J. Comput. Inf. Syst. **55**(3), 11–19 (2015). https://doi.org/10.1080/08874417.2015.11645767
52. Cheng, L., Li, Y., Li, W., Holm, E., Zhai, Q.: Understanding the violation of IS security policy in organizations: an integrated model based on social control and deterrence theory. Comput. Secur. **39**(PART B), 447–459 (2013). https://doi.org/10.1016/j.cose.2013.09.009
53. Cho, V., Ip, W.H.: A study of BYOD adoption from the lens of threat and coping appraisal of its security policy. Enterp. Inf. Syst. **12**(6), 659–673 (2018). https://doi.org/10.1080/17517575.2017.1404132
54. Compeau, D., Correia, J., Thatcher, J.B.: When constructs become obsolete: a systematic approach to evaluating and updating constructs for information systems research. MIS Q. **46**(2), 679–711 (2022). https://doi.org/10.25300/MISQ/2022/15516
55. Compeau, D., Higgins, A., Huff, S.: Social Cognitive Theory and individual reactions to computing technology: a longitudinal study. MIS Q. **23**(2), 145–158 (1999)
56. Compeau, D., Higgins, C.A.: Computer self-efficacy: development of a measure and initial test. MIS Q. **19**(2), 189 (1995). https://doi.org/10.2307/249688
57. Conway, N., Briner, R.B.: Full-time versus part-time employees: understanding the links between work status, the psychological contract, and attitudes. J. Vocat. Behav. **61**(2), 279–301 (2002)
58. Cram, A.W., D'Arcy, J., Proudfoot, J.G.: Seeing the forest and the trees: a meta-analysis of the antecedents to information security policy compliance. MIS Q.: Manag. Inf. Syst. **43**(2), 525–554 (2019). https://doi.org/10.25300/MISQ/2019/15117
59. Creswell, J.: A Concise Introduction to Mixed Methods Research. Sage Publications (2015)
60. Crossler, R.E., Bélanger, F.: Why would I use location-protective settings on my smartphone? Motivating protective behaviors and the existence of the privacy knowledge–belief gap. Inf. Syst. Res. **30**(3), 1–12 (2019). https://doi.org/10.1287/isre.2019.0846
61. Crossler, R.E., Bélanger, F.: An extended perspective on individual security behaviors: protection motivation theory and a unified security practices (USP) instrument. Data Base Adv. Inf. Syst. **45**(4), 51–71 (2014). https://doi.org/10.1145/2691517.2691521

62. Crossler, R.E., Johnston, A.C., Lowry, P.B., Hu, Q., Warkentin, M., Baskerville, R.: Future directions for behavioral information security research. Comput. Secur. **32**, 90–101 (2013). https://doi.org/10.1016/j.cose.2012.09.010
63. Crossler, R.E., Long, J.H., Loraas, T.M., Trinkle, B.S.: Understanding compliance with Bring Your Own Device policies utilizing Protection Motivation Theory: bridging the intention-behavior gap. J. Inf. Syst. **28**(1), 209–226 (2014). https://doi.org/10.2308/isys-50704
64. D'Arcy, J., Greene, G.: Security culture and the employment relationship as drivers of employees' security compliance. Inf. Manag. Comput. Secur. **22**(5), 474–489 (2014). https://doi.org/10.1108/IMCS-08-2013-0057
65. D'Arcy, J., Greene, G.: The multifaceted nature of security culture and its influence on end user behavior. In: IFIP TC 8 International Workshop on Information Systems Security Research, pp. 145–157 (2009)
66. D'Arcy, J., Hovav, A., Galletta, D.: User awareness of security countermeasures and its impact on information systems misuse: a deterrence approach. Inf. Syst. Res. **20**(1), 79–98 (2009). https://doi.org/10.1287/isre.1070.0160
67. da Veiga, A., Astakhova, L.V., Botha, A., Herselman, M.: Defining organisational information security culture—perspectives from academia and industry. Comput. Secur. **92**, 101713 (2020). https://doi.org/10.1016/j.cose.2020.101713
68. da Veiga, A., Eloff, J.H.P.: A framework and assessment instrument for information security culture. Comput. Secur. **29**(2), 196–207 (2010). https://doi.org/10.1016/j.cose.2009.09.002
69. da Veiga, A., Martins, N.: Information security culture and information protection culture: a validated assessment instrument. Comput. Law Secur. Rev. **31**(2), 243–256 (2015). https://doi.org/10.1016/j.clsr.2015.01.005
70. da Veiga, A., Martins, N.: Defining and identifying dominant information security cultures and subcultures. Comput. Secur. **70**, 72–94 (2017). https://doi.org/10.1016/j.cose.2017.05.002
71. da Veiga, A., Martins, N.: Improving the information security culture through monitoring and implementation actions illustrated through a case study. Comput. Secur. **49**, 162–176 (2015). https://doi.org/10.1016/j.cose.2014.12.006
72. Dang-Pham, D., Pittayachawan, S.: Comparing intention to avoid malware across contexts in a BYOD-enabled Australian university: a Protection Motivation Theory approach. Comput. Secur. **48**, 281–297 (2015). https://doi.org/10.1016/j.cose.2014.11.002
73. Dang-Pham, D., Pittayachawan, S., Bruno, V.: Exploring behavioral information security networks in an organizational context: an empirical case study. J. Inf. Secur. Appl. **34**, 46–62 (2017). https://doi.org/10.1016/j.jisa.2016.06.002
74. Das, A., Khan, H.U.: Security behaviors of smartphone users. Inf. Comput. Secur. **24**(1), 116–134 (2016). https://doi.org/10.1108/ICS-04-2015-0018
75. Das, T.K., Teng, B.-S.: Between trust and control: developing confidence in partner cooperation in alliances. Acad. Manag. Rev. **23**(3), 491–512 (1998). https://doi.org/10.5465/AMR.1998.926623
76. de Vargas Pinto, A., Maçada, A.C.G., Mallmann, G.L.: Workaround behaviour in information systems research. Revista de Gestão **25**(4), 430–446 (2018). https://doi.org/10.1108/rege-12-2017-0015
77. Dillard, J.P., Shen, L.: On the nature of reactance and its role in persuasive health communication. Commun. Monogr. **72**(2), 144–168 (2005). https://doi.org/10.1080/03637750500111815
78. Doargajudhur, M.S., Dell, P.: Impact of BYOD on organizational commitment: an empirical investigation. Inf. Technol. People **32**(2), 246–268 (2019). https://doi.org/10.1108/ITP-11-2017-0378
79. Doargajudhur, M.S., Dell, P.: The effect of Bring Your Own Device (BYOD) adoption on work performance and motivation. J. Comput. Inf. Syst. **60**(6), 518–529 (2018). https://doi.org/10.1080/08874417.2018.1543001
80. Donalds, C., Osei-Bryson, K.M.: Cybersecurity compliance behavior: exploring the influences of individual decision style and other antecedents. Int. J. Inf. Manag. 102056 (2019). https://doi.org/10.1016/j.ijinfomgt.2019.102056

81. Dosh, M.: Detecting insider threat within institutions using CERT dataset and different ML techniques. Period. Eng. Nat. Sci. **9**(2), 873–884 (2021). https://doi.org/10.21533/pen.v9i2.1911
82. Dzazali, S., Sulaiman, A., Zolait, A.H.: Information security landscape and maturity level: case study of Malaysian Public Service (MPS) organizations. Gov. Inf. Q. **26**(4), 584–593 (2009). https://doi.org/10.1016/j.giq.2009.04.004
83. Edwards, S.M., Li, H., Lee, J.H.: Forced exposure and psychological reactance: antecedents and consequences of the perceived intrusiveness of pop-up ads. J. Advert. **31**(3), 83–95 (2002). https://doi.org/10.1080/00913367.2002.10673678
84. Eisenhardt, K.M.: Control: Organizational and Economic Approaches **31**(2), 134–149 (1985)
85. Floyd, D.L., Prentice-Dunn, S., Rogers, R.W.: A meta-analysis of research on protection motivation. J. Appl. Soc. Psychol. **30**(2), 407–429 (2000)
86. French, A., Guo, C., Shim, J.P.: Current status, issues, and future of bring your own device (BYOD). Commun. Assoc. Inf. Syst. **35**(10)
87. Fujitsu: Can BYOD deliver real ROI ? Balancing costs and benefits in the new world of BYOD (2017)
88. Furnell, S.: Assessing website password practices—unchanged after fifteen years? Comput. Secur. **120**, 102790 (2022). https://doi.org/10.1016/j.cose.2022.102790
89. Furnell, S., Clarke, N.: Power to the people? The evolving recognition of human aspects of security. Comput. Secur. **31**(8), 983–988 (2012). https://doi.org/10.1016/j.cose.2012.08.004
90. Furnell, S., Thomson, K.L.: From culture to disobedience: recognising the varying user acceptance of IT security. Comput. Fraud Secur. **2009**(2), 5–10 (2009). https://doi.org/10.1016/S1361-3723(09)70019-3
91. Gammage, K.L., Klentrou, P.: Predicting osteoporosis prevention behaviors: health beliefs and knowledge. Am. J. Health Behav. **35**(3), 371–382 (2011). https://doi.org/10.5993/AJHB.35.3.10
92. Garba, A.B., Armarego, J., Murray, D., Kenworthy, W.: Review of the information security and privacy challenges in bring your own device (BYOD) environments. J. Inf. Privacy Secur. **11**(1), 38–54 (2015). https://doi.org/10.1080/15536548.2015.1010985
93. GCN: 8 ways to reduce unauthorized software (2014). https://gcn.com/cybersecurity/2014/11/8-ways-to-reduce-unauthorized-software/298423/
94. Georgiadou, A., Mouzakitis, S., Bounas, K., Askounis, D., Georgiadou, A.: A cyber-security culture framework for assessing organization readiness. J. Comput. Inf. Syst. 1–11 (2020). https://doi.org/10.1080/08874417.2020.1845583
95. Giwah, A.D.: Empirical Assessment of Mobile Device Users ' Information Security Behavior towards Data Breach : Leveraging Protection Motivation Theory (2019)
96. Giwah, A.D., Wang, L., Levy, Y., Hur, I.: Empirical assessment of mobile device users' information security behavior towards data breach—leveraging protection motivation theory. J. Intellect. Cap. **21**(2), 215–233 (2019). https://doi.org/10.1108/JIC-03-2019-0063
97. Guest, D.E.: Is the psychological contract worth taking seriously? J. Organ. Behav. **19**, 649–664 (1998)
98. Harris, M.A., Furnell, S., Patten, K.: Comparing the mobile device security behavior of college students and information technology professionals. J. Inf. Privacy Secur. **10**(4), 186–202 (2014). https://doi.org/10.1080/15536548.2014.974429
99. Herath, T., Rao, H.R.: Protection motivation and deterrence: a framework for security policy compliance in organisations. Eur. J. Inf. Syst. **18**(2), 106–125 (2009). https://doi.org/10.1057/ejis.2009.6
100. Herath, T., Rao, H.R.: Encouraging information security behaviors in organizations: role of penalties, pressures and perceived effectiveness. Decis. Support. Syst. **47**(2), 154–165 (2009). https://doi.org/10.1016/j.dss.2009.02.005
101. Hoh, G.C.: Head in the Clouds: how remote working behaviours are exposing organisations to cyber risks. Business Today (2020)
102. Holtsnider, B., Jaffe, B.D.: IT Manager's Handbook: Getting Your New Job Done. Elsevier Inc. (2012). https://doi.org/10.1016/b978-012370488-7/50001-4

103. Hovav, A., Putri, F.F.: This is my device! Why should I follow your rules? Employees' compliance with BYOD security policy. Pervasive Mob. Comput. **32**, 35–49 (2016). https://doi.org/10.1016/j.pmcj.2016.06.007
104. Hu, Q., Dinev, T., Hart, P., Cooke, D.: Managing employee compliance with information security policies: the critical role of top management and organizational culture. Decis. Sci. **43**(4), 615–660 (2012). https://doi.org/10.1111/j.1540-5915.2012.00361.x
105. Hu, Q., Xu, Z., Dinev, T., Ling, H.: Does deterrence work in reducing information security policy abuse by employees? Commun. ACM **54**(6), 54 (2011). https://doi.org/10.1145/1953122.1953142
106. Huang, D.L., Patrick Rau, P.L., Salvendy, G., Gao, F., Zhou, J.: Factors affecting perception of information security and their impacts on IT adoption and security practices. Int. J. Hum. Comput. Stud. **69**(12), 870–883 (2011). https://doi.org/10.1016/j.ijhcs.2011.07.007
107. Huang, E., Chuang, M.H.: Extending the theory of planned behaviour as a model to explain post-merger employee behaviour of IS use. Comput. Hum. Behav. **23**(1), 240–257 (2007). https://doi.org/10.1016/j.chb.2004.10.010
108. Hwang, I., Kim, D., Kim, T., Kim, S.: Why not comply with information security? An empirical approach for the causes of non-compliance. Online Inf. Rev. **41**(1), 2–18 (2017). https://doi.org/10.1108/OIR-11-2015-0358
109. Ifinedo, P.: Understanding information systems security policy compliance: an integration of the theory of planned behavior and the protection motivation theory. Comput. Secur. **31**(1), 83–95 (2012). https://doi.org/10.1016/j.cose.2011.10.007
110. Ifinedo, P.: Information systems security policy compliance: an empirical study of the effects of socialisation, influence, and cognition. Inf. Manag. **51**(1), 69–79 (2014). https://doi.org/10.1016/j.im.2013.10.001
111. Imgraben, J., Engelbrecht, A., Choo, K.K.R.: Always connected, but are smart mobile users getting more security savvy? A survey of smart mobile device users. Behav. Inf. Technol. **33**(12), 1347–1360 (2014). https://doi.org/10.1080/0144929X.2014.934286
112. Jarrahi, M.H., Crowston, K., Bondar, K., Katzy, B.: A pragmatic approach to managing enterprise IT infrastructures in the era of consumerization and individualization of IT. Int. J. Inf. Manag. **37**(6), 566–575 (2017). https://doi.org/10.1016/j.ijinfomgt.2017.05.016
113. Jarrahi, M.H., Nelson, S.B., Thomson, L.: Personal artifact ecologies in the context of mobile knowledge workers. Comput. Hum. Behav. **75**, 469–483 (2017). https://doi.org/10.1016/j.chb.2017.05.028
114. Jaworski, B.J.: Toward a theory of marketing control: environmental context, control types, and consequences. J. Mark. **52**(3), 23 (1988). https://doi.org/10.2307/1251447
115. Jeong, M., Lee, M., Nagesvaran, B.: Employees' use of mobile devices and their perceived outcomes in the workplace: a case of luxury hotel. Int. J. Hosp. Manag. **57**, 40–51 (2016). https://doi.org/10.1016/j.ijhm.2016.05.003
116. Johnston, A.C., Warkentin, M.: Fear appeal and information security behaviour: An empirical study **34**(1), 1–20 (2010)
117. Johnston, A.C., Warkentin, M.: Information privacy compliance in the healthcare industry. Inf. Manag. Comput. Secur. **16**(1), 5–19 (2008). https://doi.org/10.1108/09685220810862715
118. Kan, M.P.H., Fabrigar, L.R., Fishbein, M.: Theory of planned behavior. In: Encyclopedia of Personality and Individual Differences, pp. 1–8 (2017). https://doi.org/10.1007/978-3-319-28099-8
119. Kankanhalli, A., Teo, H.H., Tan, B.C.Y., Wei, K.-K.: An integrative study of information systems security effectiveness. Int. J. Inf. Manag. **23**, 139–154 (2003). https://doi.org/10.1016/S0268-4012(02)00105-6
120. Kaw, J.A., Loan, N.A., Parah, S.A., Muhammad, K., Sheikh, J.A., Bhat, G.M.: A reversible and secure patient information hiding system for IoT driven e-health. Int. J. Inf. Manag. **45**, 262–275 (2019). https://doi.org/10.1016/j.ijinfomgt.2018.09.008
121. Kearns, G.: Countering mobile device threats: a mobile device security model. J. Forens. Invest. Account. **8**(1), 157–167 (2016). http://www.nacva.com/jfia

122. Kim, A., Oh, J., Ryu, J., Lee, J., Kwon, K., Lee, K.: SoK: a systematic review of insider threat detection. J. Wirel. Mob. Netw. Ubiquitous Comput. Depend. Appl. **10**(4), 46–67 (2019). https://doi.org/10.22667/JOWUA.2019.12.31.046
123. Kirby, J.: How Cloud Computing Differs from Traditional IT Infrastructure (2020). https://micropro.com/blog/cloud-computing-vs-traditional-it-infrastructure/#:~:text=What%20is%20Traditional%20IT%20Infrastructure,a%20server%20of%20some%20sort
124. Kirsch, L.J.: The management of complex tasks in organizations: controlling the systems development process. Organ. Sci. **7**(1), 1–21 (1996). https://doi.org/10.1287/orsc.7.1.1
125. Kirsch, L.J.: Deploying common systems globally: the dynamics of control. Inf. Syst. Res. **15**(4), 374–395 (2004). https://doi.org/10.1287/isre.l040.0036
126. Knapp, K.J., Marshall, T.E., Rainer, R.K., Ford, F.N.: Information security: management's effect on culture and policy. Inf. Manag. Comput. Secur. **14**(1), 24–36 (2006). https://doi.org/10.1108/09685220610648355
127. Köffer, S., Ortbach, K.C.: Exploring the relationship between IT consumerization and job performance: a theoretical framework for future research. Commun. Assoc. Inf. Syst. **35**(1), 14 (2014)
128. Leclercq-Vandelannoitte, A.: Managing BYOD: how do organizations incorporate user-driven it innovations? Inf. Technol. People **28**(1), 2–33 (2015). https://doi.org/10.1108/ITP-11-2012-0129
129. Leclercq-Vandelannoitte, A.: Leaving employees to their own devices: new practices in the workplace. J. Bus. Strateg. **36**(5), 18–24 (2015). https://doi.org/10.1108/JBS-08-2014-0100
130. Lee, Y.: Understanding anti-plagiarism software adoption: an extended protection motivation theory perspective. Decis. Support. Syst. **50**(2), 361–369 (2011). https://doi.org/10.1016/j.dss.2010.07.009
131. Lee, Y., Larsen, K.R.: Threat or coping appraisal: determinants of SMB executives' decision to adopt anti-malware software. Eur. J. Inf. Syst. **18**(2), 177–187 (2009). https://doi.org/10.1057/ejis.2009.11
132. Li, L., He, W., Xu, L., Ash, I., Anwar, M., Yuan, X.: Investigating the impact of cybersecurity policy awareness on employees' cybersecurity behavior. Int. J. Inf. Manag. **45**, 13–24 (2019). https://doi.org/10.1016/j.ijinfomgt.2018.10.017
133. Lian, J.W.: Understanding cloud-based BYOD information security protection behaviour in smart business: in perspective of perceived value. Enterp. Inf. Syst. 1–22 (2020). https://doi.org/10.1080/17517575.2020.1791966
134. Liang, H., Xue, Y.: Avoidance of information technology threats: a theoretical perspective. MIS Q. **33**(1), 71–90 (2009)
135. Littmann, M., Moore, S.: Don't abandon security during a crisis. CISOMAG (2020). https://cisomag.eccouncil.org/dont-abandon-security-during-crisis/
136. Loch, K.D., Carr, H.H., Warkentin, M.E.: Threats to information systems: today's reality, yesterday's understanding. MIS Q. **16**(2), 173–186 (1992). https://doi.org/10.2307/249574
137. Louisnord, N.V.E.: BYOD is now standard practice, implementing it requires safe strategies. Innovation Enterprise (2017). https://channels.theinnovationenterprise.com/articles/byod-is-now-standard-practice-implementing-it-requires-safe-strategies
138. Ma, L., Ding, X., Zhang, X., Zhang, G.: Mobile users' self-disclosure behaviour on wechat: application of social cognitive theory. Mob. Inf. Syst. **2020** (2020) (Table 1). https://doi.org/10.1155/2020/8903247
139. Ma, X.: IS professionals' information security behaviors in Chinese IT organizations for information security protection. Inf. Process. Manag. **59**(1), 102744 (2022). https://doi.org/10.1016/j.ipm.2021.102744
140. Mahat, N.B., Ali, N.B.: Empowering employees through BYOD: benefits and challenges in Malaysian Public Sector. Int. J. Eng. Technol. (UAE) **7**(4), 643–649 (2018). https://doi.org/10.14419/ijet.v7i4.35.23077
141. Malaysian Communications and Multimedia Commission: Internet users survey 2020. MCMC (2020)

142. Mallmann, G.L., Maçada, A.C.G., Oliveira, M.: The influence of shadow IT usage on knowledge sharing: an exploratory study with IT users. Bus. Inf. Rev. **35**(1), 17–28 (2018). https://doi.org/10.1177/0266382118760143
143. Mani, D., Choo, K., Mubarak, S.: Information security in the South Australian real estate industry: a study of 40 real estate organisations. Inf. Manag. Comput. Secur. **22**(1), 24–41 (2014). https://doi.org/10.1108/IMCS-10-2012-0060
144. MarketWatch: Bring-Your-Own-Device (BYOD) Market Outlook—Rising Demand, Share, Trends Growth, Opportunities and Forecast 2022–2031 (2022). https://www.marketwatch.com/press-release/bring-your-own-device-byod-market-outlook---rising-demand-share-trends-growth-opportunities-and-forecast-2022-2031-2022-03-25
145. Maxis: Employee Negligence: The Greatest Mobile Threat is Sitting in Your Office (2019). https://business.maxis.com.my/en/spark/insights/employee-negligence-the-greatest-mobile-threat-is-sitting-in-your-office/
146. Mccormac, A., Parsons, K., Butavicius, M.: Preventing and Profiling Malicious Insider Attacks, pp. 1–22. Command, Control, Communications and Intelligence Division, DSTO Defence Science and Technology Organisation (2012)
147. Mejias, R.J.: An integrative model of information security awareness for assessing information systems security risk. In: Proceedings of the Annual Hawaii International Conference on System Sciences, pp. 3258–3267 (2012). https://doi.org/10.1109/HICSS.2012.104
148. Mansfield-Devine, S.: Interview: BYOD and the enterprise network. Computer Fraud Security **2012**(4), 14–17. https://doi.org/10.1016/S1361-3723(12)70031-3
149. Merhi, M.I., Ahluwalia, P.: Examining the impact of deterrence factors and norms on resistance to Information Systems Security. Comput. Hum. Behav. **92**, 37–46 (2019). https://doi.org/10.1016/j.chb.2018.10.031
150. Miles, A.D.: A taxonomy of research gaps: identifying and defining the seven research gaps. J. Res. Methods Strateg. **2017**, 1–15 (2017). https://www.academia.edu/35505149/RESEARCH_A_Taxonomy_of_Research_Gaps_Identifying_and_Defining_the_Seven_Research_Gaps
151. Mimiaga, M.J., Reisner, S.L., Reilly, L., Soroudi, N., Safren, S.A.: Individual interventions. HIV Prevention, pp. 203–239 (2009). https://doi.org/10.1016/B978-0-12-374235-3.00008-X
152. Moody, G.D., Siponen, M., Pahnila, S.: Toward a unified model of information security policy compliance. MIS Q. **42**(1), 285–311 (2018). https://doi.org/10.25300/MISQ/2018/13853
153. Müller-Bloch, C., Kranz, J.: A framework for rigorously identifying research gaps in qualitative literature reviews. In: 2015 International Conference on Information Systems: Exploring the Information Frontier, ICIS 2015, pp. 1–19 (2015)
154. Müller, J.: Smartphone penetration rate as share of the population in Malaysia from 2015 to 2025. Statistica, Aug (2021)
155. Muncaster, P.: #COVID19 Drives Phishing Emails Up 667% in Under a Month. Infosecurity Magazine (2020). https://www.infosecurity-magazine.com/news/covid19-drive-phishing-emails-667/
156. Musarurwa, S., Gamundani, A.M., Shava, F.B.: An assessment of BYOD control in higher learning institutions: a Namibian perspective. In: Cunningham, P., Cunningham, M. (eds.) 2019 IST-Africa Week Conference (IST-Africa), pp. 1–9. IEEE (2019). https://doi.org/10.23919/istafrica.2019.8764853
157. Mustafa, S.Z., Kar, A.K.: Prioritization of multi-dimensional risk for digital services using the generalized analytic network process. Digit. Policy Reg. Gov. **21**(2), 146–163 (2019). https://doi.org/10.1108/DPRG-06-2018-0031
158. Mylonas, A., Kastania, A., Gritzalis, D.: Delegate the smartphone user? Security awareness in smartphone platforms. Comput. Secur. **34**, 47–66 (2013). https://doi.org/10.1016/j.cose.2012.11.004
159. Naidoo, R.: A multi-level influence model of COVID-19 themed cybercrime. Eur. J. Inf. Syst. **29**(3), 306–321 (2020). https://doi.org/10.1080/0960085X.2020.1771222
160. Nasir, A., Arshah, R.A., Hamid, M.R.A., Fahmy, S.: An analysis on the dimensions of information security culture concept: a review. J. Inf. Secur. Appl. **44**, 12–22 (2019). https://doi.org/10.1016/j.jisa.2018.11.003

161. National Security Council: Malaysia Cyber Security Strategy 2020–2024, pp. 1–90 (2020)
162. Nel, F., Drevin, L.: Key elements of an information security culture in organisations. Inf. Comput. Secur. **27**(2), 146–164 (2019). https://doi.org/10.1108/ICS-12-2016-0095
163. Ng, B.Y., Kankanhalli, A., Xu, Y. (Calvin).: Studying users' computer security behavior: a health belief perspective. Decis. Support Syst. **46**(4), 815–825 (2009). https://doi.org/10.1016/j.dss.2008.11.010
164. Niehaves, B., Köffer, S., Ortbach, K.: IT consumerization—a theory and practice review. In: 18th Americas Conference on Information Systems 2012, AMCIS 2012, vol. 6, pp. 4705–4713 (2012)
165. NordPass: The Misfortunate Passwords of Fortune 500 Companies. NordPass (2021). https://nordpass.com/fortune-500-password-study/
166. Oberheide, J., Jahanian, F.:. When mobile is harder than fixed (and vice versa): demystifying security challenges in mobile environments. In: HotMobile 2010: The 11th Workshop on Mobile Computing Systems and Applications, pp. 43–48 (2010). https://doi.org/10.1145/1734583.1734595
167. Ogbanufe, O., Pavur, R.: Going through the emotions of regret and fear: revisiting protection motivation for identity theft protection. Int. J. Inf. Manag. **62**, 102432 (2022). https://doi.org/10.1016/j.ijinfomgt.2021.102432
168. Ouchi, W.G.: The relationship between organizational structure and organizational control author(s). Adm. Sci. Q. **22**(1), 95–113 (1977)
169. Ouchi, W.G.: A conceptual framework for the design of Organizational Control mechanisms. Manag. Sci. **25**(9), 833–848 (1979). https://doi.org/10.1287/mnsc.25.9.833
170. Ouchi, W.G., Maguire, M.A.: Organizational control: two functions. Adm. Sci. Q. **20**(4), 559 (1975). https://doi.org/10.2307/2392023
171. Paananen, H., Lapke, M., Siponen, M.: State of the art in information security policy development. Comput. Secur. **88**, 101608 (2020). https://doi.org/10.1016/j.cose.2019.101608
172. Padayachee, K.: Taxonomy of compliant information security behavior. Comput. Secur. **31**(5), 673–680 (2012). https://doi.org/10.1016/j.cose.2012.04.004
173. Pahnila, S., Siponen, M., Mahmood, A.: Employees' behavior towards IS security policy compliance. In: Proceedings of the Annual Hawaii International Conference on System Sciences, pp. 1–10 (2007). https://doi.org/10.1109/HICSS.2007.206
174. Palanisamy, R., Norman, A.A., Kiah, M.L.M.: Compliance with bring your own device security policies in organizations: a systematic literature review. Comput. Secur. 101998 (2020). https://doi.org/10.1016/j.cose.2020.101998
175. Pfleeger, C.P., Pfleeger, S.L., Margylies, J.: Security in Computing, 5th edn. Prentice Hall (2015). http://ptgmedia.pearsoncmg.com/images/9780134085043/samplepages/9780134085043.pdf
176. Plotnikoff, R.C., Trinh, L., Courneya, K.S., Karunamuni, N., Sigal, R.J.: Predictors of physical activity in adults with type 2 diabetes. Am. J. Health Behav. **35**(3), 359–370 (2011). https://doi.org/10.5993/AJHB.35.3.9
177. Ponemon Institute: 2020 Cost of Insider Threats Global Report (2020)
178. Posey, C., Roberts, T.L., Lowry, P.B.: The impact of organizational commitment on insiders motivation to protect organizational information assets. J. Manag. Inf. Syst. **32**(4), 179–214 (2015). https://doi.org/10.1080/07421222.2015.1138374
179. Posey, C., Roberts, T.L., Lowry, P.B., Bennett, R.J., Courtney, J.F.: Insiders' protection of organizational information assets: development of a systematics-based taxonomy and Theory Of Diversity for Protection-Motivated behaviors. MIS Q. **37**(4), 1189–1210 (2013)
180. Posey, C., Roberts, T.L., Lowry, P.B., Hightower, R.T.: Bridging the divide: a qualitative comparison of information security thought patterns between information security professionals and ordinary organizational insiders. Inf. Manag. **51**(5), 551–567 (2014). https://doi.org/10.1016/j.im.2014.03.009
181. Poulter, D.R., Chapman, P., Bibby, P.A., Clarke, D.D., Crundall, D.: An application of the theory of planned behaviour to truck driving behaviour and compliance with regulations. Accid. Anal. Prev. **40**(6), 2058–2064 (2008). https://doi.org/10.1016/j.aap.2008.09.002

182. Prentice-Dunn, S., Rogers, R.W.: Protection motivation theory and preventive health: beyond the health belief model. Health Educ. Res. **1**(3), 153–161. http://her.oxfordjournals.org/
183. Pu, W., Li, S., Bott, G.J., Esposito, M., Thatcher, J.B.: To disclose or not to disclose: an evaluation of the effects of information control and social network transparency. Comput. Secur. **112**, 102509 (2022). https://doi.org/10.1016/j.cose.2021.102509
184. Rho, H.O., Ryu, I.: The impact of information technology threat on IT appropriation and IT avoidance. In: Proceedings—3rd International Conference on Computational Intelligence, Communication Systems and Networks, CICSyN 2011, pp. 195–200 (2011). https://doi.org/10.1109/CICSyN.2011.50
185. Robinson, K.A., Saldanha, I.J., McKoy, N.A.: Development of a framework to identify research gaps from systematic reviews. J. Clin. Epidemiol. **64**(12), 1325–1330 (2011). https://doi.org/10.1016/j.jclinepi.2011.06.009
186. Rogers, R.W.: A protection motivation theory of fear appeals and attitude change. J. Psychol. **91**(1), 93–114 (1975). https://doi.org/10.1080/00223980.1975.9915803
187. Romer, H.: Best practices for BYOD security. Comput. Fraud Secur. **1**, 13–15 (2014). https://doi.org/10.1016/S1361-3723(14)70007-7
188. Rosenstock, I.M.: Historical origins of the health belief model. Health Educ. Behav. **2**(4), 328–335 (1974). https://doi.org/10.1177/109019817400200403
189. Rousseau, D.M.: Psychological and implied contracts in organizations. Empl. Responsib. Rights J. **2**, 121–139 (1989)
190. Ruighaver, A.B., Maynard, S.B., Chang, S.: Organisational security culture: extending the end-user perspective. Comput. Secur. **26**(1), 56–62 (2007). https://doi.org/10.1016/j.cose.2006.10.008
191. Sabtu, S.B.M., Mohamad, K.M.: Critical information infrastructure protection framework development: preliminary findings from the malaysian public sector. In: 2020 8th International Conference on Information Technology and Multimedia, ICIMU 2020 (2020). https://doi.org/10.1109/ICIMU49871.2020.9263397
192. Safa, N.S., Sookhak, M., Von Solms, R., Furnell, S., Ghani, N.A., Herawan, T.: Information security conscious care behaviour formation in organizations. Comput. Secur. **53**, 65–78 (2015). https://doi.org/10.1016/j.cose.2015.05.012
193. Safa, N.S., Von Solms, R.: An information security knowledge sharing model in organizations. Comput. Hum. Behav. **57**, 442–451 (2016). https://doi.org/10.1016/j.chb.2015.12.037
194. Sallehudin, H., Aman, A.H.M., Razak, R.C., Ismail, M., Bakar, N.A.A., Fadzil, A.F.M., Baker, R.: Performance and key factors of cloud computing implementation in the public sector. Int. J. Bus. Soc. **21**(1), 134–152 (2020)
195. Schein, E.H.: Organizational culture and leadership. Jossey Bass, San Fransissco (1985). https://doi.org/10.1016/0024-6301(93)90120-5
196. Schlienger, T., Teufel, S.: Analyzing information security culture: increased trust by an appropriate information security culture. In: Proceedings—International Workshop on Database and Expert Systems Applications, DEXA, 2003-Janua (September 2018), pp. 405–409 (2003). https://doi.org/10.1109/DEXA.2003.1232055
197. Schlienger, T., Teufel, S.: Information security culture—from analysis to change. South Afr. Comput. J. **31**, 46–52 (2003)
198. Sebescen, N., Vitak, J.: Securing the human: employee security vulnerability risk in organizational settings. J. Am. Soc. Inf. Sci. **68**(9), 2237–2247 (2017). https://doi.org/10.1002/asi
199. Silic, M., Back, A.: Shadow IT—a view from behind the curtain. Comput. Secur. **45**, 274–283 (2014). https://doi.org/10.1016/j.cose.2014.06.007
200. Siponen, M., Pahnila, S., Mahmood, A.: Factors Influencing Protection Motivation and IS Security Policy Compliance (2006)
201. Siponen, M., Vance, A.: Neutralization: new insights into the problem of employee Information Systems security policy violations. MIS Q. **34**(3), 487–502 (2010)
202. Sniehotta, F.F., Presseau, J., Araújo-Soares, V.: Time to retire the theory of planned behaviour. Health Psychol. Rev. **8**(1), 1–7 (2014). https://doi.org/10.1080/17437199.2013.869710

203. Sommestad, T., Hallberg, J., Lundholm, K., Bengtsson, J.: Variables influencing information security policy compliance: a systematic review of quantitative studies. Inf. Manag. Comput. Secur. **22**(1), 42–75 (2014). https://doi.org/10.1108/IMCS-08-2012-0045
204. Soomro, Z.A., Shah, M.H., Ahmed, J.: Information security management needs more holistic approach: a literature review. Int. J. Inf. Manag. **36**(2), 215–225 (2016). https://doi.org/10.1016/j.ijinfomgt.2015.11.009
205. Souppaya, M., Scarfone, K.: Guidelines for managing the security of mobile devices in the enterprise (2013). https://doi.org/10.6028/NIST.SP.800-124r1
206. Stagliano, T., Dipoalo, A., Coonelly, P.: Consumerization of IT. Mathematics and Computer Science Capstones, p. 10 (2013). https://digitalcommons.lasalle.edu/mathcompcapstones/10
207. Stephens, K.K., Ford, J.L.: Unintended consequences of a strategically ambiguous organizational policy selectively restricting mobile device use at work. Mob. Media Commun. **4**(2), 186–204 (2016). https://doi.org/10.1177/2050157915619211
208. Tanenbaum, W.A.: IT systems put security into health care cybersecurity. J. Health Care Compliance 21–27 (2016)
209. Tariq, Q.: Online Dangers During the Pandemic. The Star Malaysia, 14 Sept (2020). https://www.pressreader.com/malaysia/the-star-malaysia-star2/20200914/281913070528015
210. Taylor, S., Todd, P.: Decomposition and crossover effects in the theory of planned behavior: a study of consumer adoption intentions. Int. J. Res. Mark. **12**(2), 137–155 (1995). https://doi.org/10.1016/0167-8116(94)00019-K
211. Teoh, C.S., Kamil Mahmood, A., Dzazali, S.: Cyber security challenges in organisations: a case study in Malaysia. In: 2018 4th International Conference on Computer and Information Sciences: Revolutionising Digital Landscape for Sustainable Smart Society, ICCOINS 2018—Proceedings, pp. 1–6 (2018). https://doi.org/10.1109/ICCOINS.2018.8510569
212. The CERT Insider Threat Team: Unintentional insider threats: a foundational study. Carnegie Mellon University, August (2013). https://doi.org/10.1155/2014/106363
213. Theoharidou, M., Mylonas, A., Gritzalis, D.: A risk assessment method for smartphones. In: IFIP International Information Security Conference, pp. 443–456 (2012)
214. Thompson, N., McGill, T.J., Wang, X.: "Security begins at home": determinants of home computer and mobile device security behavior. Comput. Secur. **70**, 376–391 (2017). https://doi.org/10.1016/j.cose.2017.07.003
215. Trang, S., Plics, P., Masuch, K., Hengstler, S.: Fostering information security compliance: comparing the predictive power of social learning theory and deterrence theory. Bandura **1977**, 1–10 (2019)
216. Tsohou, A., Holtkamp, P.: Are users competent to comply with information security policies? An analysis of professional competence models. Inf. Technol. People **31**(5), 1047–1068 (2018). https://doi.org/10.1108/ITP-02-2017-0052
217. Tsohou, A., Karyda, M., Kokolakis, S.: Analyzing the role of cognitive and cultural biases in the internalization of information security policies: recommendations for information security awareness programs. Comput. Secur. **52**, 128–141 (2015). https://doi.org/10.1016/j.cose.2015.04.006
218. Tu, Z., Adkins, J., Zhao, G.Y.: Complying with BYOD security policies: a moderation model based on Protection Motivation Theory. J. Midwest Assoc. Inf. Syst. **2019**(1), 11–28 (2019). https://doi.org/10.17705/3jmwa.000045
219. Tu, Z., Yuan, Y.: Understanding user's behaviors in coping with security threat of mobile devices loss and theft. In: Proceedings of the Annual Hawaii International Conference on System Sciences, pp. 1393–1402 (2012). https://doi.org/10.1109/HICSS.2012.620
220. Van Niekerk, J.F., Von Solms, R.: Information security culture: a management perspective. Comput. Secur. **29**(4), 476–486 (2010). https://doi.org/10.1016/j.cose.2009.10.005
221. Vance, A.: Why do employees violate is security policies? Insights from multiple theoretical perspectives. Doctoral dissertation, University of Oulu, Finland (2010)
222. Vance, A., Siponen, M., Pahnila, S.: Motivating IS security compliance: insights from habit and Protection Motivation Theory. Inf. Manag. **49**, 190–198 (2012). https://doi.org/10.1016/j.im.2012.04.002

223. Venkatesh, V., Brown, S.A., Bala, H.: Bridging the qualitative-quantitative divide: guidelines for conducting mixed methods research in information systems. Manag. Inf. Syst. Q. **37**(1), 21–54 (2013). https://doi.org/10.25300/MISQ/2013/37.1.02
224. Verkijika, S.F.: Understanding smartphone security behaviors: an extension of the protection motivation theory with anticipated regret. Comput. Secur. **77**, 860–870 (2018). https://doi.org/10.1016/j.cose.2018.03.008
225. Vorakulpipat, C., Sirapaisan, S., Rattanalerdnusorn, E., Savangsuk, V.: A policy-based framework for preserving confidentiality in BYOD environments: a review of information security perspectives. Secur. Commun. Netw. **2017**, 1–11 (2017). https://doi.org/10.1155/2017/2057260
226. Vrhovec, S., Markelj, B.: Relating mobile device use and adherence to information security policy with data breach consequences in hospitals. J. Univ. Comput. Sci. **24**(5), 634–645 (2018)
227. Wang, X., Xu, J.: Deterrence and leadership factors: which are important for information security policy compliance in the hotel industry. Tour. Manag. **84**, 104282 (2021). https://doi.org/10.1016/j.tourman.2021.104282
228. Wani, T.A., Mendoza, A., Gray, K.: BYOD usage and security behaviour of hospital clinical staff : an Australian survey. Int. J. Med. Inform. 104839 (2022). https://doi.org/10.1016/j.ijmedinf.2022.104839
229. Wani, T.A., Mendoza, A., Gray, K., Smolenaers, F.: Status of bring-your-own-device (BYOD) security practices in Australian hospitals—a national survey. Health Policy Technol. 100627 (2022). https://doi.org/10.1016/j.hlpt.2022.100627
230. Weeger, A., Wang, X., Gewald, H.:. IT consumerization: BYOD-program acceptance and its impact on employer attractiveness. J. Comput. Inf. Syst. **56**(1) (2016). https://doi.org/10.1080/08874417.2015.11645795
231. Weeger, A., Wang, X., Gewald, H., Raisinghani, M., Otavio, S., Grant, G., Siddhi, P.: Determinants of intention to participate in corporate BYOD-programs: the case of digital natives. Inf. Syst. Front. 1–42 (2018). https://doi.org/10.13140/RG.2.1.5160.0726
232. Weeger, A., Wang, X., Gewald, H., Raisinghani, M., Sanchez, O., Grant, G., Pittayachawan, S.: Determinants of intention to participate in corporate BYOD-programs: the case of digital natives. Inf. Syst. Front. **22**(1), 203–219 (2020). https://doi.org/10.1007/s10796-018-9857-4
233. Weiß, F., Leimeister, J.M.: Consumerization: IT innovations from the consumer market as a challenge for corporate IT. Bus. Inf. Syst. Eng. **4**(6), 363–366 (2012). https://doi.org/10.1007/s12599-012-0234-4
234. White, T., Lao, T.: IT managers' and IT professionals' mobile device security strategies: a qualitative exploratorary case study. J. Manag. Policy Pract. **22**(3), 90–100 (2020). https://doi.org/10.33423/jmpp.v22i3.4684
235. Whitman, M.E., Mattord, H.J.:Principles of Information Security, 6th edn. Cengage Learning (2018)
236. Wiley, A., McCormac, A., Calic, D.: More than the individual: examining the relationship between culture and information security awareness. Comput. Secur. **88** (2020). https://doi.org/10.1016/j.cose.2019.101640
237. Woon, I.M.Y., Kankanhalli, A.: Investigation of IS professionals' intention to practise secure development of applications. Int. J. Hum. Comput. Stud. **65**(1), 29–41 (2007). https://doi.org/10.1016/j.ijhcs.2006.08.003
238. Workman, M.: A field study of corporate employee monitoring: attitudes, absenteeism, and the moderating influences of procedural justice perceptions. Inf. Organ. **19**(4), 218–232 (2009). https://doi.org/10.1016/j.infoandorg.2009.06.001
239. Wu, D., Moody, G.D., Zhang, J., Lowry, P.B.: Effects of the design of mobile security notifications and mobile app usability on users' security perceptions and continued use intention. Inf. Manag. **57**(5), 103235 (2020). https://doi.org/10.1016/j.im.2019.103235
240. Zahadat, N.: Mobile Security: A Systems Engineering Framework for Implementing Bring Your Own Device (BYOD) Security Through the Combination of Policy Management and Technology. ProQuest Dissertations and Theses, pp. 1–202 (2016). http://search.proquest.

com.library.capella.edu/docview/1769079590?accountid=27965%5Cnhttp://wv9lq5ld3p.search.serialssolutions.com.library.capella.edu/?ctx_ver=Z39.88-2004&ctx_enc=info:ofi/enc:UTF-8&rfr_id=info:sid/ProQuest+Dissertations+&+Theses+Global&r
241. Zahadat, N., Blessner, P., Blackburn, T., Olson, B.A.: BYOD security engineering: a framework and its analysis. Comput. Secur. **55**, 81–99 (2015). https://doi.org/10.1016/j.cose.2015.06.011
242. Zakaria, O.: Internalisation of information security culture amongst employees through basic security knowledge. In: IFIP International Federation for Information Processing, vol. 201, pp. 437–441 (2006). https://doi.org/10.1007/0-387-33406-8_38

Chapter 3
Theoretical Framework and Hypotheses Development

Abstract The previous chapter has reviewed theories and factors thoroughly to pave the way for the discovery of important factors impacting the BYOD security policy compliance intention. The identified factors serve as the basis for the conceptualization of the BYOD security policy compliance research framework. Detailed reviews of each hypothesized independent and dependent variables are drawn in order to justify the deployment of the variables. This chapter discusses the theories used to guide this research and describes the integration of Organizational Control Theory, Social Cognitive Theory and Security culture and how it can be applied to the BYOD security policy compliance. The hypotheses are presented in the following sections with a conclusion at the end of the chapter.

The previous chapter has reviewed theories and factors thoroughly to pave the way for the discovery of important factors impacting the BYOD security policy compliance intention. The identified factors serve as the basis for the conceptualization of the BYOD security policy compliance research framework. Detailed reviews of each hypothesized independent and dependent variables are drawn in order to justify the deployment of the variables. This chapter discusses the theories used to guide this research and describes the integration of Organizational Control Theory, Social Cognitive Theory and Security culture and how it can be applied to the BYOD security policy compliance. The hypotheses are presented in the following sections with a conclusion at the end of the chapter.

3.1 Integrating Organizational Control Theory, Social Cognitive Theory and Security Culture

The aim of this research is to establish a BYOD security policy framework and empirically identify the framework's significant factors in predicting the BYOD security policy compliance intention among CNII agencies' employees.

© The Author(s), under exclusive license to Springer Nature Switzerland AG 2025
R. Palanisamy et al., *Bring Your Own Device Security Policy Compliance Framework*, Information Systems Engineering and Management 37,
https://doi.org/10.1007/978-3-031-86801-6_3

The purpose of proposing the integrated theoretical framework, is to address the identified research problem presented in Chap. 1 and research gaps as explained in Sect. 2.9. The proposed factors in the conceptual framework are formed based on past research's significant works and described in the following sections.

3.2 Elements of Control in BYOD Security Policy (Specification of Policy, Evaluation of Compliance and Perceived Mandatoriness)

A good security policy is essential for improving an organization's BYOD environment [18, 19, 34, 64, 123, 135, 146]. Every organization must design a BYOD security policy and privacy strategies that safeguards organizational information and guides employees' security behavior [54, 64, 137, 146]. It is crucial as the BYOD security policy prevents the lack of direction that can cause insecure practices in the organization due to misinformation [138].

Controls comprise of three elements: specification, evaluation and reward or penalty [46, 82]. Specification in information security refers to an organization's security policies and procedures that govern the behaviors of personnel in order to achieve a particular result [22]. It translates into specific and measurable objectives, in order for the employees to adhere to the information system security requirements [53, 115]. Employees are informed of the significance of compliance to the organization, as well as the organization's requirements, and this has an effect on employee attitudes [35]. These factors are important as security specification are put in place to encourage compliant behavior [128]. An organization establishes rules to encourage employees to act in ways that safeguard information assets and avoid security breaches through their behavior. To begin, a policy should be carefully defined in order to avoid ambiguity and misinterpretation, especially among non-management personnel. Second, the policy should be implemented gradually, as abrupt restrictions on normal mobile device habits can jeopardize the connection between employers and employees. Employee and third-party access, control, and protection of BYOD resources must be clearly established in the security policy [19]. In addition to the policies, it is crucial for businesses to describe how the settings are implemented and where they are located [18]. A security policy should also be evaluated on a regular basis to verify that the BYOD security and privacy controls, as well as the policies that incorporate dynamic BYOD practices, are in place [19, 146] can serve as a constant reminder about the security compliance requirements to employees.

Evaluation refers to the assessment of an individual's behavior in following the organization's security policies and procedures. It refers to the implementation of a feedback system on the output through auditing and performance evaluations [22, 122, 126]. Evaluation of behavior is performed to ensure employees behave in adherence to the security policy and validate compliance with the specified security measure. Evaluation focuses on what is measured to ensure positive outcomes are achieved [95]. The examination of data is used to determine if an employee has executed the required security practices and procedures in accordance with the security policies [82]. Organization can monitor their employees either by physically monitoring the employees or by having systems to measure them. These factors are crucial for employees to be aware of the intended actions and adhere to the policies indicated. Besides, evaluation provides feedback and learning opportunities to employees on their behavior and security requirement [22, 35]. Without proper evaluation, a security policy is meaningless [22].

The requirement to adhere to the security policy is mandatory, where participation is not optional or where compliance is required by the organization. According to a study conducted by [22] involving 1698 valid respondents from all levels of employees, the specification of policy and evaluation of compliance determine the perceived mandatoriness. Therefore, organizations should provide employees with precise information regarding their security procedures to minimize ambiguity [34, 123]. This can establish a positive environment and self-enforcement. Besides, it can also inform employees of the benefits of policy compliance [135].

3.3 Elements of SCT in BYOD Security Policy Compliance (BYOD IT Support and Self-efficacy)

3.3.1 BYOD IT Support

Support is the availability of support for individuals to utilize their personal devices. SCT views support as one of the characteristics that positively influence self-efficacy. The traditional help desk has evolved well beyond PC support. Smartphones, laptops and tablet computers are now considered essential tools for employees. On a monthly basis, organizations receive requests for mobile device support from 10 to 12% of users on average [29]. IT support can involve technical issues like how to deal with connectivity issues, configuration of applications, and email issues. When employees are using their devices to improve work productivity, organizations are committed to provide technical support as it is impactful to the revenue of the organization, besides ensuring secure use of device [58]. The provision of support for employees who need assistance with their devices will encourage them to adhere to the organization's security rules. The provision of support to employees who need support with the use of their devices will inspire them to adhere to the organization's security requirements [114]. Having prompt, competent, and engaging IT support can be advantageous for employees [30], which can have a positive impact on their self-efficacy. Employees who perceive that they are receiving assistance will improve their ability and perceptions of their capacity to manage a certain task [50]. When employees are confident, they can execute actions prescribed in the security policies

[64]. When employees are using their devices to improve work productivity, organizations are committed to provide technical support as it is impactful to the revenue of the organization, besides ensuring secure use of device [58].

IT support can involve technical issues like how to deal with connectivity issues, configuration of applications, and email issues. Strong IT support is required to implement a successful and secure BYOD program in the organization.

3.3.2 *Self-efficacy*

Self-efficacy is "the conviction that one can successfully execute the behavior required to produce outcomes" [15]. The concept of self-efficacy plays a central role in Social Cognitive Theory in which the belief that a person is capable of using a certain system influences significantly his/her ability to perform certain tasks [24]. Bandura [16] projected that self-efficacy expectations would influence individuals' actual capacity to do the behavior. Similarly, the idea asserts that individuals with low self-efficacy in a certain task tend to avoid that task. Self-efficacy, or a person's confidence in his or her ability to act and persist despite the challenges in using technology [113, 131].

The concept of self-efficacy has been used to explain individual computer related behaviors [32, 33]. Compeau and Higgins [33] discovered that self-efficacy exerts a considerable influence on an individual's performance with computer use, thus, those with higher self-efficacy about IS use are more likely to utilize such systems at work than individuals with lower self-efficacy. Extending Social Cognitive Theory to the setting of information security, individuals' beliefs about their abilities to secure their information and information systems may help to explain security practices and their intention to continue their existing efforts [107, 143]. Self-efficacy has been utilized regularly to evaluate employees' compliance with the organization's security policy. In the field of information security, self-efficacy is described as an individual's confidence in their ability to secure information against threats such as unauthorized disclosure, alteration, and loss. Additionally, self-efficacy has been regularly utilized to assess employees' adherence to the organization's security policy. In the field of information security, self-efficacy is described as an individual's belief in their ability to secure information against dangers such as unauthorized disclosure, alteration, and loss [107]. Previous research has demonstrated that self-efficacy has a substantial impact on an employee's intention to adhere to the organization's security policy. Previous research has demonstrated that self-efficacy has a significant impact on an employee's intention to adhere to the organization's security policy [22, 23, 25, 60, 77, 87]. Furthermore, self-efficacy has also been reported to impact actual behavior [97, 142], observed behavior [141] and intentions to perform behaviors [69].

Employees who believe that he or she has the ability to act in accordance with the policies is likely to have to comply with policies [60]. Chan et al. [25] examined the causal role of employee self-efficacy and IS security behavior in IT intensive organizations. This study indicated that employees' belief in their efficacy in IS security influenced their decision to perform or not perform IS security related task, particularly those required by the security policies. Rhee et al. [107] proposed that self-efficacy in information security enhances employees' security practice, strengthens security efforts and security care behavior [88]. Clearly, it has been found as a key predictor of security policy compliance because it correlates with employees' confidence in their ability to conduct the needed security behavior. In this study, self-efficacy is related to the employees' belief on their ability to interact and secure their devices. Self-efficacy is important because if employees are uneasy with some technologies and find them difficult to implement, therefore they will not practice them even though they can prevent security threats.

3.4 Elements of Security Culture in BYOD Security Policy Compliance (Top Management Commitment, Device Monitoring)

3.4.1 Top Management

The full support of the organization's management is crucial because it ensures information security effectiveness and establishes a secure environment for information handling [66, 109]. The management commitment refers to documenting organization ISPs, ensuring that policies and procedures are put in practice, providing security training and awareness programs [84]. They are also responsible for encouraging and monitoring employees' adherence towards the security policies [68].

If an organization wants to get its mission and security policy to align well together, it needs a dedicated management team [91]. Sebescen and Vitak [112] advocated that an organization's security culture and atmosphere can be improved via open and honest communication between upper management and lower-level employees. This is essential for establishing and maintaining a robust and secure environment.

Top management must be involved in information security and take on responsibility for security management and the implementation of security measures. The commitment of senior management translates into moral and financial support for the proper implementation of information security measures [55, 78]. They communicate the significance of information security to the rest of the organization and enable employees to alter their approach to information security [47]. Top management has also been posited as an important influence on information security perceptions, beliefs and attitudes [67], but its effects are rarely examined empirically [35].

According to AlKalbani et al. [6] and Liu, et al. [90] senior management commitment could take the form of providing active support through firm choices, investments, and activities taken to enforce the organization-wide information security strategy. Apart from this, active engagement in addressing the organization's information security challenges and alignment of the information security policy's goals with other organizational goals will strengthen employees' belief in the importance of complying with information security policies and procedures. Regardless of the prominence of social pressures, strong management commitment to enforcing information security measures would result in a higher level of information security compliance inside public enterprises. Managers can intervene in corporate culture to trigger changes in attitudes towards more compliant and secure employee behaviors and it is possible by preparing them with proper training programs.

Puhakainen and Siponen [106] found that visible top management support (e.g., actively advocating security issues and leading by example through their own compliance behavior) influenced employee information security attitudes and was crucial for achieving employee information security policy compliance. Kankanhalli et al. [78] discovered that organizations with stronger support from upper management engaged in more preventive initiatives than those with weaker support. Humaidi and Balakrishnan [68] studied the indirect effects of management support on the compliance behavior of healthcare employees towards information security policies (ISPs). Their findings showed that top management support combined with other constructs from the Theory of Planned Behaviour were significant predictors of user compliance behavior towards information security policies. D'Arcy and Greene [36] explored three specific dimensions of information security culture, namely top management commitment, security communication and computer monitoring. According to the study, all three constructs contributed to security policy adherence. A recent study by Ameen et al. [10] discovered that top management support had a significant effect on behavioral intention.

In addition to this, top management can have an impact on IT initiatives at the organizational level by supporting in the definition of the mission, goals, and strategy, as well as serving as a facilitator of IT usage [26, 65] besides contributing to the organizations' compliance culture [28, 70].

3.4.2 *Device Monitoring*

Organizations have deployed computer monitoring to enhance compliance with rules and regulations, hence decreasing security breaches. Monitoring strategies span from employee security behavior evaluation to electronic surveillance of computer, email, and Internet usage [124]. Mobile device management (MDM) platforms allow organizations to monitor and control practically all operations of employee devices. Text, voice, and data usage, location, phone condition, and device status may all be monitored. The system can disable hardware functions such as cameras, Bluetooth, and GPS to exert control over the hardware. Apps are approved and required for devices, and users' ability to install blacklisted apps is restricted. MDM systems can safeguard an organization's data by remotely wiping data from devices that are no longer under its control (e.g., lost device and employee termination).

In some professions, employee monitoring has been utilized to assess job performance. Today, computers and the Internet are indispensable corporate tools, and monitoring has expanded beyond a focus on task performance to encompass all electronic use. Accessing device information and limiting device usage enables employers to electronically monitor their personnel. Monitoring improves the organization's ability to hold employees accountable for their electronic activity, increase staff efficiency and effectiveness, and minimize Internet abuse [89].

D'Arcy et al. [37] discovered that when computer monitoring is known to be in place, IS events are drastically reduced. When employees are observed, they are more likely to adhere to the security policy. While, there are plenty of literature providing rich understanding of monitoring in mandatory settings, it fails to address volitional conditions.

3.5 Security Education Training and Awareness (SETA)

Training plays a crucial role in creating awareness among employees on BYOD security risk [18, 39, 91, 98, 112, 133, 146]. The purpose of continual training programs is to induce positive behavioral changes in employees and improve their risk awareness. Changes can be made by educating employees on how to improve their security behavior [14, 139]. In this regard, organizations must ensure that their employees are aware of the risks associated with using any online platform to share information [18]. Training programs would teach employees to be cautious of device vulnerabilities and how to safeguard them. Additionally, through these training sessions, organizations should also communicate their security and privacy issues as well as their principles on security [19].

ISP compliance and preserving the organizations' information security necessitate that employee have a comprehensive awareness of the policies and processes governing information security compliance in their organizations [130, 143, 144]. These studies also found that information security knowledge has a significant impact on information security compliance. According to Siponen [23], Bulgurcu et al. [117] offering security knowledge is the most significant component in convincing employees to improve their security policy compliance behaviors.

Education and training can advance a person's self-efficacy by increasing their understanding of how to prevent or minimize a security issue. The majority of prior studies concluded that security incidents are caused by employees' inability to perform security tasks. According to Chen and Li [27], having a variety of security training and education programs can be an effective way to strengthen the security of mobile devices. Son [120] noted that an organization can establish training and education programs and strengthen the link between the objective of the IS policy and its personnel.

Liao et al. [89] recommends management to invest in training programs, so that it can enhance employees' human capital, employees' knowledge, skills and abilities in order for them to play their role as a line of defense. In the context of IS security, SETA programs emphasize providing personnel with general understanding of the IS security environment and the necessary skills to fulfill the required IS security tasks.

Additionally, Tu et al. [132] stressed the importance of mandatory trainings or organizational incentives offered to employees who attend one. Making them read and sign a document admitting their grasp of the policy is not necessarily indicative of their familiarity with and comprehension of the security policies. Along with that, the study also found that threat and coping evaluation are crucial in determining compliance behavior and these factors should be considered while building training programs.

Creating a fear appeal by explaining the repercussions of poor security behavior or providing real-world examples of security events [112] might increase employees' concerns, leading to a decline in information sharing behavior via mobile devices [18]. The survey indicated that rising privacy concerns are a factor in preventing employees from sharing information online and may eventually inspire staff to adopt mobile device security settings. Verkijika [133] observed that exposing users to the negative repercussions of their behavior through training programs may cause them to reflect regret over their earlier behavior. Verkijika [133] and Mylonas et al. [98] also discovered that users with technical expertise were more likely to examine security alerts and make proper security decisions for permission to install configurations, hence reducing security vulnerabilities.

On the other hand, Zahadat et al. [146] suggested that training modules be revised periodically to ensure that they include all security-related topics. This facilitates a comprehensive grasp of security topics [112]. Further, they underlined that training must be researched in depth, and that corporations must view this as mandatory in order to transform employees into assets. Social learning should also be promoted because employees learn from their surroundings through observation, imitation, and social cues [131]. In this regard, training will ensure that employees are oriented towards positive security compliance behavior while gaining a sense of the procedures and mechanisms that need to be implemented, consequently securing business data.

From a different angle, Dang-Pham and Pittayachawan [39] discovered that training may not be the only factor that contributes to self-efficacy. They hypothesized that it may be attributable to a group phenomenon in which professional support and informal assistance from coworkers contribute to employees' sense of self-efficacy. A community of practice that shares its expertise and experiences on information security should be established in tandem with a series of training sessions, rather than as a one-time, ineffective effort.

In the context of the adoption of the BYOD security policy, it appears that when there is limited or no training for mobile devices, employees would likely seek assistance from their social surroundings [94, 132]. This indicates the significance of the social environment that may influence employees' decisions regarding security compliance behavior.

With mobile devices, users make their own security decisions. Leaving control solely in the hands of staff poses security risks. This feature which distinguishes BYOD from traditional computing, emphasizes the need for security knowledge and training. Hence, trainings should incorporate elements of technology-specific competencies. Users are frequently uninformed of the risks these gadgets provide, and even if they are aware to some extent, they still need to know how to mitigate them [7, 71].

According to previous studies, SETA is crucial for enhancing cybersecurity culture and safeguarding firms against information security events resulting from employee irresponsibility [9, 101]. The standard ISO/IEC 27002:2013 (2013) stipulates that SETA must be integrated into the organization's overall information security management system. The standard requires that organizations offer all personnel with sufficient education, training, and awareness in security procedures and proper use of information systems [72].

3.6 Psychological Ownership

It is defined as a mental state in which individuals feel as though a target of ownership is theirs [105]. Psychological ownership consists of both cognitive and affective processes [104]. This feeling can be developed in tangible objects such as property and also intangible objects such as concepts, innovative activities. An individual's intellectual perspective represents his or her knowledge, beliefs, and opinions about the target of ownership, and it is created in conjunction with an emotional attachment to the target. When feelings of personal ownership are threatened by an external entity, the emotional aspect of ownership might manifest significantly [103]. In terms of self-identification, one of the cognitive processes connected with psychological ownership is viewing the target of ownership as an extension of oneself and as a component of one's identity [43]. Psychological ownership has three fundamental purposes: self-identity, efficacy/effectivity, and a sense of location [103].

Prior research on psychological ownership has established that psychological ownership is a primary determinant of behaviors that affect the target of ownership [43, 105]. It has been demonstrated that psychological ownership has good consequences on a person's attitude. A sense of ownership also inspires feelings of responsibilities toward the target [45]. Researchers have also proven that when a person experiences a strong sense of oneness with an objective, that person's sense of responsibility for that target is similarly heightened [13]. The owner will be more inclined to defend the privileges associated with the target's ownership. Psychological ownership is associated with security policy compliance as increasing levels of psychological ownership with mobile devices should motivate employees to comply. It triggers a sense of responsibility for the target [94]. This is particularly relevant in the context of this research. However, research incorporating psychological ownership in information security contexts has been limited [12, 31, 80, 83, 127, 145], therefore this construct may offer similar potential for expanding the area of research in the realm of information security, specifically for BYOD security policy compliance.

3.7 Proposed Research Framework and Hypotheses

This study put forth an integrated theoretical model to fill in the gaps in the literature and make significant contributions to the body of knowledge. The extensive literature evaluations and research gaps described in earlier sections led to the construction of the research framework for this study. Organizational Control Theory (OCT), Social Cognitive Theory (SCT) were selected as the central theoretical lenses along with security culture, SETA, and psychological ownership. Figure 3.1 depicts the proposed research framework for this study. In the next sections, hypotheses are presented formally and literature-based arguments supporting their relevance are given.

3.7.1 *Specification of Security Policy*

Statement is defined as a formal specification of a necessary behavior or outcome and the arrangements made to gather data on compliance with the specified behavior or

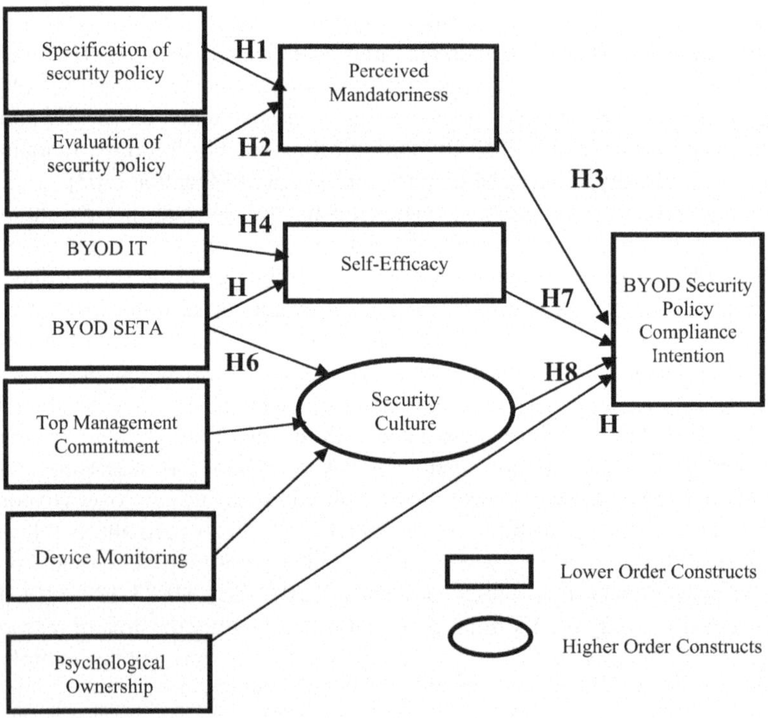

Fig. 3.1 Proposed research framework

outcome. In order to practice control, specification of desired behavior or outcomes is vital in the form of documented procedures [46, 82]. Individuals will perceive that they are required to comply if rules, policies, and processes are specified. Security policy should be carefully drafted to achieve the targeted goals and objectives [53, 55]. Understandable policies give a clear instruction to the employees to achieve the desired behavior. Having a policy, but if the employees do not comprehend it, it will be unsuccessful in addressing security behavior [2] because if they are poorly informed, they will be unable to participate in the security process [49]. Having employees understand the security policy, organizations' expectations and the importance of compliance is one of the most cost-effective ways to reduce security risks [41, 56, 102, 117]. This study will include specification of security policy as this may be seen as mandatory by employees, hence the following hypothesis is proposed:

H1: Specification will be positively associated with the employees' perceived mandatoriness of the established set of security policies.

3.7.2 *Evaluation of Security Policy*

Evaluation is crucial for determining an individual's conformance with specified behaviors or outcomes [82]. Employees will disregard policies if management never evaluates compliance, according to [22]. Some evaluation is performed manually by examining specific individual machines and log files. By combining human observation, impromptu checks, and log data analysis, management can determine whether or not personnel adhere to security policies and procedures [22]. Importantly, evaluation of security policy demonstrates to all employees that compliance is mandatory [22], its importance and organizations expectations on the employees [35]. Regularly, managers should review the behaviors of their staff until they adopt secure conduct as stipulated by the security policies [5]. In the study by D'Arcy et al. [37], it was determined that monitoring methods are an effective countermeasure for minimizing non-compliant behavior by convincing users that misused activities will be detected. Therefore, the following hypothesis is proposed:

H2: Evaluation of security policy compliance will be positively associated with the employees' perceived mandatoriness of the established ICT security policy.

3.7.3 Perceived Mandatoriness in Information Security

Boss et al. [22] in their study defined mandatoriness as the "degree to which individuals perceive that compliance with existing security policies and procedures is compulsory or expected by organizational management". A control system needs to be implemented to observe and evaluate individual behavior against some standard [99, 100]. A mandate from management in regards to policy can persuade individuals to follow the code of practice for information security. Therefore, if the employees perceive the code of practice based on ISO27002/2013 standard is mandatory, they are more likely practice security behavior and hence comply with the BYOD security policy. The mandatory organizational rules, guidelines and requirements in the security policy is crucial in shaping employees' behavior [17]. In the workplace, policies are enforced either by technology or by the threat of punishment, and not by a voluntary milieu in which employees exercise their free will in complying with security policies [96]. The administrative approach to information security has been primarily structured on legal regulations, standards such as ISO17799, documented policies and procedures for individual and organizational behavior, control and monitoring, and distribution of privileges. The policies and processes that control employee conduct constitute corporate governance [111].

H3: Individual's perceived mandatoriness of compliance is positively associated with BYOD security policy compliance.

3.7.4 BYOD IT Support

SCT identifies support as one of the characteristics that positively influence self-efficacy [33, 50]. According to the study by Compeau and Higgins [33], when organizations provide computer support to individuals who require it, it improves both their actual and perceived ability to accomplish a particular task. In this context, the availability of an IT support team to address technical difficulties pertaining to employees' personal devices may influence their sense of self-efficacy and outcome expectations, hence influencing their compliance behavior. IT support is typical in environments where all resources are owned by the organization, however this may not be the case in situations where personal devices are utilized [64]. The availability of IT support may also bolster employee confidence in their ability to carry out the BYOD security policy's mandated tasks.

H3: H4: BYOD IT support is positively associated with self-efficacy.

3.7.5 BYOD SETA

SETA involves an understanding of security threats and their consequences, an understanding of security threats and their consequences, concerns and risks and knowledge about the cost of security problems [23]. The widespread vulnerabilities and security attacks are caused by lack of awareness of security measures [8, 20]. Study also found that employees themselves feel that there is a lack of proper security awareness in managing personal devices' data [125]. Employees cannot be vigilant against threats or perform security policy compliance behavior if they lack awareness and knowledge. The BYOD ISP requires employees to perform specific task; such as setting a password or PIN, running antivirus and so on, these tasks require certain level of knowledge and skill [129]. Besides, they would fail to evaluate how threatening the situation is and deploy the resources they have to handle the threats [74]. SETA is an element that is crucial for employees need to protect themselves from the huge threat landscape and to reduce security breaches [48, 51, 59]. Studies shows that individuals are not aware of the risk pertaining to the use of smartphones leading to lack of privacy protective behavior [7]. Similarly, lack of knowledge has been validated as a factor for intention to violate security policy [119]. Providing security awareness is the most important factor in influencing employees with their compliance behaviors [106, 117]. According to [75], creating SETA is the initial step in motivating employees' compliance. SETA will raise the visibility of threats and lower the costs associated with handling security breaches. Organizations have to focus on information security awareness programs, rather than just relying on security policies alone [52].

H5: SETA is positively associated with self-efficacy.
H6: SETA is positively associated with security culture.

3.7.6 Self-efficacy

Self-efficacy is a crucial element in SCT that is defined as the belief that one can successfully execute the behavior required to produce outcomes [15]. It has been commonly acknowledged that the concept of self-efficacy plays a central role in SCT in which the belief that a person is capable of using a certain system influences significantly his/her performance [16, 24]. It is the individuals' belief regarding how easy and convenient it would be for them to use the security controls [3, 76, 77]. Bandura [16] predicted that self-efficacy expectations will influence individuals' actual ability to perform the behavior. It is their belief in abilities to gather motivation, organize resources and actions needed to exercise control. Rhee et al. [107] in their study, use self-efficacy to study security behavior in two aspects, the adoption of security technology and security conscious care behavior of using computer and

Internet. Since its initial conceptualization, a number of studies have applied the concept of self-efficacy to explain individual computer related behaviors [32, 33] and people's security behaviors [11, 22, 60, 77]. It is one of the key aspects of human activity especially in their interaction with technology and security [85]. In most cases, increased self-efficacy is directly related with more favorable intentions or behaviors [44]. Hence, this leads to the following hypothesis:

H7: Self-efficacy is positively associated with BYOD security policy compliance.

3.7.7 *Security Culture*

There are various opinions and definitions of security culture, however it can be viewed as shared beliefs, norms, and values, as well as the manner in which individuals work inside an organization. Building an IS culture in organizations is the first step towards establishing compliance behavior among employees [2]. Effective information security culture is characterized by the incorporation of security norms and practices into the organization's assumptions and core values. Security culture is more effective in organizations that prioritizes information integrity and confidentiality through the organizations' philosophies and management. Security culture is able to minimize the risks to IT assets as well as employees' interactions with the assets whether intentional or unintentional [62]. Research has shown that security culture is an important factor in increasing compliant behavior [121]. Despite the paucity of studies addressing the relationship between culture and security behavior, the existing literature indicates that management support and dedication is essential [25, 36, 84] is crucial in developing security culture. Security culture research is still in its infancy, according to [38, 79], and its role should be researched further. The culture of the organization impacts the employees behavior towards information security [136].

3.7.8 *Top Management*

Security culture is a multi-dimensional concept whereby top management is regarded as the main construct for this factor as their importance have been widely explained in literature [134]. According to research, the views that decision-makers inside businesses have about security, as well as the quality of various security processes, are distinctive aspects of a security culture. Top management commitment and involvement are critical in building a security culture, and it will encourage staff to fulfil their information security obligations [4, 40, 114]. In their study, Hu et al. [66] found that top management support influences employee compliance with security policies. It is also seen as a critical success factor for implementing information security since more financial and technical resources are likely to be available for IS with good top management support [66, 78, 84]. Top management involvement translates policies into actions [57], since it strengthens employees' links to the firm and fosters a positive culture, which encourages employees to follow security policies. Involvement of top management transforms policies into actions. It increases employees' ties to the organization and creates a positive culture, which increases security policy compliance among employees.

3.7.9 Device Monitoring

Computer monitoring is the another dimension of security culture that is often used by organizations to gain compliance with rules and regulations [37]. Among activities involved for computer monitoring are tracking employees' Internet use, recording network activities, and performing security audits [1]. Employees' devices can be monitored when the devices connect to the organization wireless network [110]. Monitoring can be used for oversighting behavior of employees to ensure data integrity and confidentiality is protected [63, 73]. Monitoring employees with information security controls is important to detect undesirable behaviour and prevent its consequences [92]. It helps to identify employees who engage in information security misbehavior. When employees realize that their behaviour will be detected, they are more likely to avoid noncompliance behaviour [23] (Trang et al. 2019). By increasing the information security surveillance, it reduces delinquent behavior and promotes compliant employees [108].

H8: Security culture is positively associated with BYOD security policy compliance.

3.7.10 Psychological Ownership

Psychological ownership is a situation in which individuals feel a possessive bond with something they perceive as theirs [103]. It is the interaction between a person and an object in which the item is perceived to be linked to the self [103]. The user feels ownership of hardware and the software and information that it contains, such that it becomes an 'extension of the self' [93]. Increased psychological ownership of a target might result in heightened feelings of responsibilities toward it. It was found to enhance a number of desirable behaviors, which comprise increased effort in care and maintenance [103]. It can trigger defensive and protective behavior

[31]. Psychological ownership of one's mobile device was found to be significantly related to intention to perform security behavior among home computer users [11]. Thompson et al. [127] extended this factor to the BYOD setting and found psychological ownership to influence individuals' security behavior. In the context of personal devices, this involves the feelings of ownership for the devices, software and the information that resides in it [94]. Employees purchase their own personal devices, which have become the focal point of their lives, as they use them to contact with family and friends, listen to music, watch movies, check email, shop online, and even set their alarms for work. It has become an inseparable part of employees' daily lives. However, more research is required to explore its potential role especially in BYOD workplace setting [21] as personal devices like smartphones have become integral part of an individual's self-identity and such phenomenon is affecting people's daily lives decisions and behavior. Therefore, it is important to investigate how this centrality.

H9: Psychological ownership is positively associated with BYOD security policy compliance.

3.7.11 *BYOD Security Policy Compliance Intention*

Intention is an important element in the formation of behavior [108] as it shows the assurance from the employees on carrying out their responsibilities. Intention is viewed as a prerequisite for an actual behavior of individuals [86, 116, 118, 119, 140]. ISP compliance activities are a series of activities that comply with managerial and technical information, which are written statements on the roles and responsibilities employees must fulfil in order to protect important information in their organization [42, 61, 81].

3.8 Summary

This chapter explained the constructs used in developing the research framework based on findings of SLR. The proposed research framework will be explored further with security experts' perception of factors that influence security policy compliance. Following that, validity of the framework and the hypotheses put forth will be validated via quantitative research methods. Both qualitative and quantitative results are discussed in Chap. 5.

References

1. Ahmad, Z., Ong, T.S., Liew, T.H., Norhashim, M.: Security monitoring and information security assurance behaviour among employees: an empirical analysis. Inf. Comput. Secur. **27**(2), 165–188 (2019). https://doi.org/10.1108/ICS-10-2017-0073
2. Akande, A.O.: Policy awareness , enforcement , and maintenance: a comparative quantitative approach to information security effectiveness in a Bring Your Own Device (BYOD) environment
3. Al-Omari, A., El-Gayar, O., Deokar, A.: Security policy compliance: user acceptance perspective. In: Proceedings of the Annual Hawaii International Conference on System Sciences, pp. 3317–3326 (2012). https://doi.org/10.1109/HICSS.2012.516
4. AlGhamdi, S., Win, K.T., Vlahu-Gjorgievska, E.: Information security governance challenges and critical success factors: systematic review. Comput. Secur. **99**, 102030 (2020). https://doi.org/10.1016/j.cose.2020.102030
5. Ali, R.F., Dominic, P.D.D., Ali, S.E.A., Rehman, M., Sohail, A.: Information security behavior and information security policy compliance: a systematic literature review for identifying the transformation process from noncompliance to compliance. Appl. Sci. (Switzerland), **11**(8) (2021). https://doi.org/10.3390/app11083383
6. AlKalbani, A., Deng, H., Kam, B.: Organisational Security Culture and information security compliance for E-Government development: the moderating effect of social pressure. In: 19th Pacific Asia Conference on Information Systems, PACIS 2015, vol. 65 (2015). http://aisel.aisnet.org/pacis2015/65
7. Allam, S., Flowerday, S.V., Flowerday, E.: Smartphone information security awareness: a victim of operational pressures. Comput. Secur. **42**, 56–65 (2014). https://doi.org/10.1016/j.cose.2014.01.005
8. Alsaleh, M., Alomar, N., Alarifi, A.: Smartphone users: understanding how security mechanisms are perceived and new persuasive methods. PLoS ONE **12**(3) (2017). https://doi.org/10.1371/journal.pone.0173284
9. Alshaikh, M.: Developing cybersecurity culture to influence employees behavior: a practice perspective. Comput. Secur. 102003 (2020). https://doi.org/10.1016/j.cose.2020.102003
10. Ameen, N., Tarhini, A., Shah, M.H., Madichie, N., Paul, J., Choudrie, J.: Keeping customers' data secure: a cross-cultural study of cybersecurity compliance among the Gen-Mobile workforce. Comput. Hum. Behav. **114** (2021). https://doi.org/10.1016/j.chb.2020.106531
11. Anderson, C.L., Agarwal, R.: Practicing safe computing: a multimethod empirical examination of home computer user security behavioral intentions. MIS Q. **34**(3), 613–643 (2010)
12. Anderson, C.L., Agarwal, R.: Practicing safe computing: a multimethod empirical examination of home computer user security behavioral intentions: appendix. MIS Q. **86**, 1282–1288 (2010). https://doi.org/10.1016/j.jmpt.2006.06.011
13. Avey, J.B., Avolio, B.J., Crossley, C.D., Luthans, F.: Psychological ownership: theoretical extensions, measurement and relation to work outcomes. J. Organ. Behav. **30**(2), 173–191 (2009). https://doi.org/10.1002/job.583
14. Balapour, A., Nikkhah, H.R., Sabherwal, R.: Mobile application security: role of perceived privacy as the predictor of security perceptions. Int. J. Inf. Manag. **52**, 102063 (2020). https://doi.org/10.1016/j.ijinfomgt.2019.102063
15. Bandura, A.: Self-efficacy: toward a unifying theory of behavioral change. Psychol. Rev. **84**(2), 191–215 (1977). https://doi.org/10.1037/0033-295X.84.2.191
16. Bandura, A.: Social Foundations of Thought and Action. Prentice Hall (1986)
17. Bauer, S., Bernroider, E.W.N.: From information security awareness to reasoned compliant action. ACM SIGMIS Database: DATABASE Adv. Inf. Syst. **48**(3), 44–68 (2017). https://doi.org/10.1145/3130515.3130519
18. Belanger, F., Crossler, R.E.: Dealing with digital traces: understanding protective behaviors on mobile devices. J. Strat. Inf. Syst. **28**(1), 34–49 (2018). https://doi.org/10.1016/j.jsis.2018.11.002

19. Bello, A.G., Murray, D., Armarego, J.: A systematic approach to investigating how information security and privacy can be achieved in BYOD environments. Inf. Comput. Secur. **25**(4), 475–492 (2017). https://doi.org/10.1108/ICS-03-2016-0025
20. Bitton, R., Finkelshtein, A., Sidi, L., Puzis, R., Rokach, L., Shabtai, A.: Taxonomy of mobile users' security awareness. Comput. Secur. **73**, 266–293 (2018). https://doi.org/10.1016/j.cose.2017.10.015
21. Blythe, J.M., Coventry, L.: Costly but effective: comparing the factors that influence employee anti-malware behaviours. Comput. Hum. Behav. **87**, 87–97 (2018). https://doi.org/10.1016/j.chb.2018.05.023
22. Boss, S.R., Kirsch, L.J., Angermeier, I., Shingler, R.A., Boss, R.W.: If someone is watching, I'll do what I'm asked: mandatoriness, control, and information security. Eur. J. Inf. Syst. **18**(2), 151–164 (2009). https://doi.org/10.1057/ejis.2009.8
23. Bulgurcu, B., Cavusoglu, H., Benbasat, I.: Information security policy compliance: an empirical study of rationality-based beliefs and information security awareness. MIS Q. **34**(3), 523–548 (2010). https://doi.org/10.1093/bja/aeq366
24. Carillo, K.D.A.: Social cognitive theory in is research–literature review, criticism, and research agenda. In: International Conference on Information Systems, Technology and Management, pp. 20–31 (2010). https://doi.org/10.1007/978-3-642-29166-1
25. Chan, M., Woon, I., Kankanhalli, A.: Perceptions of information security in the workplace: linking information security climate to compliant behavior. J. Inf. Privacy Secur. **1**(3), 18–41 (2005). https://doi.org/10.1080/15536548.2005.10855772
26. Chatterjee, D., Grewal, R., Sambamurthy, V.: Shaping up for E-commerce: institutional enablers of the organizational assimilation of web technologies. MIS Q. **26**(2), 65–89 (2002)
27. Chen, H., Li, W.: Mobile device users' privacy security assurance behaviour: a technology threat avoidance perspective. Inf. Comput. Secur. **25**(3), 330–344 (2017). https://doi.org/10.1108/ICS-04-2016-0027
28. Chen, X., Wu, D., Chen, L., Teng, J.K.L.: Sanction severity and employees' information security policy compliance: investigating mediating, moderating, and control variables. Inf. Manag. **55**(8), 1049–1060 (2018). https://doi.org/10.1016/j.im.2018.05.011
29. Cherkala, R.: The Importance of Employee-Centric Mobile Support. Vox Mobile (2013)
30. Chowdhury, N.H., Adam, M.T.P., Skinner, G.: The impact of time pressure on cybersecurity behaviour: a systematic literature review. Behav. Inf. Technol. **38**(12), 1290–1308 (2019). https://doi.org/10.1080/0144929X.2019.1583769
31. Cichy, P., Salge, T.O., Kohli, R.: Extending the privacy calculus: the role of psychological ownership. In: 35th International Conference on Information Systems "Building a Better World Through Information Systems", ICIS 2014 (2014)
32. Compeau, D., Higgins, A., Huff, S.: Social Cognitive Theory and individual reactions to computing technology: a longitudinal study. MIS Q. **23**(2), 145–158 (1999)
33. Compeau, D., Higgins, C.A.: Computer self-efficacy: development of a measure and initial test. MIS Q. **19**(2), 189 (1995). https://doi.org/10.2307/249688
34. Crossler, R.E., Long, J.H., Loraas, T.M., Trinkle, B.S.: Understanding compliance with Bring Your Own Device policies utilizing Protection Motivation Theory: bridging the intention-behavior gap. J. Inf. Syst. **28**(1), 209–226 (2014). https://doi.org/10.2308/isys-50704
35. Cuganesan, S., Steele, C., Hart, A.: How senior management and workplace norms influence information security attitudes and self-efficacy. Behav. Inf. Technol. **37**(1), 50–65 (2018). https://doi.org/10.1080/0144929X.2017.1397193
36. D'Arcy, J., Greene, G.: Security culture and the employment relationship as drivers of employees' security compliance. Inf. Manag. Comput. Secur. **22**(5), 474–489 (2014). https://doi.org/10.1108/IMCS-08-2013-0057
37. D'Arcy, J., Hovav, A., Galletta, D.: User awareness of security countermeasures and its impact on information systems misuse: a deterrence approach. Inf. Syst. Res. **20**(1), 79–98 (2009). https://doi.org/10.1287/isre.1070.0160
38. da Veiga, A., Astakhova, L.V., Botha, A., Herselman, M.: Defining organisational information security culture—perspectives from academia and industry. Comput. Secur. **92**, 101713 (2020). https://doi.org/10.1016/j.cose.2020.101713

39. Dang-Pham, D., Pittayachawan, S.: Comparing intention to avoid malware across contexts in a BYOD-enabled Australian university: a Protection Motivation Theory approach. Comput. Secur. **48**, 281–297 (2015). https://doi.org/10.1016/j.cose.2014.11.002
40. Daud, M.: National Information Infrastructure Organisations and Cyber Security Compliance in Malaysia. Doctoral thesis, University of Malaya (2018)
41. Dhillon, G.: Managing and controlling computer misuse. Inf. Manag. Comput. Secur. **7**(4), 171–175 (1999). https://doi.org/10.1007/978-1-349-14454-9_6
42. Dhillon, G., Backhouse, J.: Current directions in IS security research: towards socio-organizational perspectives. Inf. Syst. J. **11**(2), 127–153 (2001). https://doi.org/10.1046/j.1365-2575.2001.00099.x
43. Dittmar, H.: The social psychology of material possessions: To have is to be (1992)
44. Donalds, C., Osei-Bryson, K.M.: Cybersecurity compliance behavior: exploring the influences of individual decision style and other antecedents. Int. J. Inf. Manag. 102056 (2019). https://doi.org/10.1016/j.ijinfomgt.2019.102056
45. Dyne, L.V., Pierce, J.L.: Psychological ownership and feelings of possession: three field studies predicting employee attitudes and organizational citizenship behavior. J. Organ. Behav. **25**(4), 439–459 (2004). http://www.jstor.org/stable/4093721
46. Eisenhardt, K.M.: Control: organizational and economic approaches. Manag. Sci. **31**(2), 134–149 (1985)
47. Flores, W.R., Ekstedt, M.: Shaping intention to resist social engineering through transformational leadership, information security culture and awareness. Comput. Secur. **59**, 26–44 (2016). https://doi.org/10.1016/j.cose.2016.01.004
48. Furnell, S., Clarke, N.: Power to the people? The evolving recognition of human aspects of security. Comput. Secur. **31**(8), 983–988 (2012). https://doi.org/10.1016/j.cose.2012.08.004
49. Furnell, S., Khern-am-nuai, W., Esmael, R., Yang, W., Li, N.: Enhancing security behaviour by supporting the user. Comput. Secur. **75**, 1–9 (2018). https://doi.org/10.1016/j.cose.2018.01.016
50. Galvez, S.M., Shackman, J.D., Guzman, I.R., Ho, S.M.: Factors affecting individual information security practices. In: SIGMIS-CPR'15, June, vol. 26(8), pp. 555–557 (2015). https://doi.org/10.1145/2751957.2751966
51. Garcia-Perez, A., Sallos, M.P., Tiwasing, P.: Dimensions of cybersecurity performance and crisis response in critical infrastructure organisations: an intellectual capital perspective. J. Intellect. Cap. (2021). https://doi.org/10.1108/JIC-06-2021-0166
52. Giwah, A.D.: Empirical Assessment of Mobile Device Users ' Information Security Behavior Towards Data Breach: Leveraging Protection Motivation Theory (2019)
53. Goel, S., Chengalur-Smith, I.N.: Metrics for characterizing the form of security policies. J. Strateg. Inf. Syst. **19**(4), 281–295 (2010). https://doi.org/10.1016/j.jsis.2010.10.002
54. Goh, C.H., Teoh, A.P.: Determining Bring Your Own Device (BYOD) security policy compliance among Malaysian teleworkers: perceived cybersecurity governance as moderator. In: 2021 IEEE 5th International Conference on Information Technology, Information Systems and Electrical Engineering (ICITISEE), pp. 305–310 (2021)
55. Goo, J., Yim, M.S., Kim, D.J.: A path to successful management of employee security compliance: an empirical study of information security climate. IEEE Trans. Prof. Commun. **57**(4), 286–308 (2014). https://doi.org/10.1109/TPC.2014.2374011
56. Hagen, J.M., Albrechtsen, E., Hovden, J.: Implementation and effectiveness of organizational information security measures. Inf. Manag. Comput. Secur. **16**(4), 377–397 (2008). https://doi.org/10.1108/09685220810908796
57. Hassandoust, F., Subasinghage, M., Johnston, A.C.: Information & Management: a neo-institutional perspective on the establishment of information security knowledge sharing practices. Inf. Manag. **59**(1), 103574 (2022). https://doi.org/10.1016/j.im.2021.103574
58. Hatton, S.: Device Security Measures You Can Take for Your Organization. Endsight (2022). https://www.endsight.net/blog/mobile-device-security-for-your-organization
59. He, W., Ash, I., Anwar, M., Li, L., Yuan, X., Xu, L., Tian, X.: Improving employees' intellectual capacity for cybersecurity through evidence-based malware training. J. Intellect. Cap. **21**(2), 203–213 (2020). https://doi.org/10.1108/JIC-05-2019-0112

60. Herath, T., Rao, H.R.: Protection motivation and deterrence: a framework for security policy compliance in organisations. Eur. J. Inf. Syst. **18**(2), 106–125 (2009). https://doi.org/10.1057/ejis.2009.6
61. Herath, T., Rao, H.R.: Encouraging information security behaviors in organizations: role of penalties, pressures and perceived effectiveness. Decis. Support. Syst. **47**(2), 154–165 (2009). https://doi.org/10.1016/j.dss.2009.02.005
62. Hina, S., Dominic, P.D.D.: Information security policies' compliance: a perspective for higher education institutions. J. Comput. Inf. Syst. **60**(3), 201–211 (2018). https://doi.org/10.1080/08874417.2018.1432996
63. Hina, S., Selvam, D.D.D.P., Lowry, P.B.: Institutional governance and protection motivation: theoretical insights into shaping employees' security compliance behavior in higher education institutions in the developing world. Comput. Secur. **87**, 101594 (2019). https://doi.org/10.1016/j.cose.2019.101594
64. Hovav, A., Putri, F.F.: This is my device! Why should I follow your rules? Employees' compliance with BYOD security policy. Pervasive Mob. Comput. **32**, 35–49 (2016). https://doi.org/10.1016/j.pmcj.2016.06.007
65. Hsia, T.L., Chiang, A.J., Wu, J.H., Teng, N.N.H., Rubin, A.D.: What drives E-Health usage? Integrated institutional forces and top management perspectives. Comput. Hum. Behav. **97**, 260–270 (2019). https://doi.org/10.1016/j.chb.2019.01.010
66. Hu, Q., Dinev, T., Hart, P., Cooke, D.: Managing employee compliance with information security policies: the critical role of top management and organizational culture. Decis. Sci. **43**(4), 615–660 (2012). https://doi.org/10.1111/j.1540-5915.2012.00361.x
67. Hu, Q., Hart, P., Cooke, D.: The role of external and internal influences on information systems security—a neo-institutional perspective. J. Strateg. Inf. Syst. **16**(2), 153–172 (2007). https://doi.org/10.1016/j.jsis.2007.05.004
68. Humaidi, N., Balakrishnan, V.: Leadership styles and information security compliance behavior: the mediator effect of information security awareness. Int. J. Inf. Educ. Technol. **5**(4), 311–318 (2015). https://doi.org/10.7763/IJIET.2015.V5.522
69. Ifinedo, P.: Understanding information systems security policy compliance: an integration of the theory of planned behavior and the protection motivation theory. Comput. Secur. **31**(1), 83–95 (2012). https://doi.org/10.1016/j.cose.2011.10.007
70. Ifinedo, P.: Critical times for organizations: what should be done to curb workers' noncompliance with IS security policy guidelines? Inf. Syst. Manag. **33**(1), 30–41 (2016). https://doi.org/10.1080/10580530.2015.1117868
71. Imgraben, J., Engelbrecht, A., Choo, K.K.R.: Always connected, but are smart mobile users getting more security savvy? A survey of smart mobile device users. Behav. Inf. Technol. **33**(12), 1347–1360 (2014). https://doi.org/10.1080/0144929X.2014.934286
72. International Organization for Standardization: ISO/IEC 27001:2013—Information technology—Security techniques—Information security management systems—Requirements. ISO (2013)
73. Jackson, G.W., Rahman, S.S.M.: Security governance, management and strategic alignment via capabilities. In: 2017 International Conference on Computational Science and Computational Intelligence, pp. 44–49 (2017). https://doi.org/10.1109/CSCI.2017.8
74. Jaeger, L., Eckhardt, A.: Eyes wide open: the role of situational information security awareness for security-related behaviour. Inf. Syst. J. 1–44 (2020). https://doi.org/10.1111/isj.12317
75. Jaeger, L., Eckhardt, A., Kroenung, J.: The role of deterrability for the effect of multi-level sanctions on information security policy compliance: results of a multi-group analysis. Inf. Manag. minor revision, 103318 (2020). https://doi.org/10.1016/j.im.2020.103318
76. Johnston, A.C., Warkentin, M.: Fear appeal and information security behaviour: an empirical study. MIS Q. **34**(1), 1–20 (2010)
77. Johnston, A.C., Warkentin, M., Siponen, M.: An enhanced fear appeal rhetorical framework: leveraging threats to the human asset through sanctioning rhetoric. MIS Q.: Manag. Inf. Syst. **39**(1), 113–134 (2015). https://doi.org/10.25300/MISQ/2015/39.1.06

78. Kankanhalli, A., Teo, H.H., Tan, B.C.Y., Wei, K.-K.: An integrative study of information systems security effectiveness. Int. J. Inf. Manag. **23**, 139–154 (2003). https://doi.org/10.1016/S0268-4012(02)00105-6
79. Karlsson, F., Åström, J., Karlsson, M.: Information security culture state-of-the-art review between 2000 and 2013. Inf. Comput. Secur. **23**(3), 246–285 (2015). https://doi.org/10.1108/ICS-05-2014-0033
80. Kehr, F., Wentzel, D., Kowatsch, T.: Privacy paradox revised: pre-existing attitudes, psychological ownership, and actual disclosure. In: 35th International Conference on Information Systems "Building a Better World Through Information Systems", ICIS 2014, vol. 4, pp. 1–12 (2014)
81. Kirsch, L., Boss, S.: The last line of defense: motivating employees to follow corporate security guidelines. In: Twenty Eighth International Conference on Information Systems (2007). http://aisel.aisnet.org/icis2007
82. Kirsch, L.J.: Deploying common systems globally: the dynamics of control. Inf. Syst. Res. **15**(4), 374–395 (2004). https://doi.org/10.1287/isre.1040.0036
83. Klesel, M., Ndicu, M., Niehaves, B.: Exploring psychological ownership of IT: an empirical study. In: 24th European Conference on Information Systems, ECIS 2016 (2016)
84. Knapp, K.J., Marshall, T.E., Rainer, R.K., Ford, F.N.: Information security: management's effect on culture and policy. Inf. Manag. Comput. Secur. **14**(1), 24–36 (2006). https://doi.org/10.1108/09685220610648355
85. Latikka, R., Turja, T., Oksanen, A.: Self-efficacy and acceptance of robots. Comput. Hum. Behav. **93**, 157–163 (2019). https://doi.org/10.1016/j.chb.2018.12.017
86. Lebek, B., Uffen, J., Neumann, M., Hohler, B., Reitner, M.H.: Information security awareness and behavior: a theory-based literature review. Manag. Res. Rev. **37**(12), 1049–1092 (2014). https://doi.org/10.1108/MRR-04-2013-0085
87. Li, L., He, W., Xu, L., Ash, I., Anwar, M., Yuan, X.: Investigating the impact of cybersecurity policy awareness on employees' cybersecurity behavior. Int. J. Inf. Manag. **45**, 13–24 (2019). https://doi.org/10.1016/j.ijinfomgt.2018.10.017
88. Lian, J.W.: Understanding cloud-based BYOD information security protection behaviour in smart business: in perspective of perceived value. Enterp. Inf. Syst. 1–22 (2020). https://doi.org/10.1080/17517575.2020.1791966
89. Liao, Q., Gurung, A., Luo, X.I.N., Long, L.I.: Workplace management and employee misuse: does punishment matter? J. Comput. Inf. Syst. **50**(2), 49–59 (2009). https://doi.org/10.1080/08874417.2009.11645384
90. Liu, G.H.W., Wang, E., Huang Chua, C.E.: Leveraging social capital to obtain top management support in complex, cross-functional it projects. J. Assoc. Inf. Syst. **16**(8), 707–737 (2015). https://doi.org/10.17705/1jais.00404
91. Mani, D., Choo, K., Mubarak, S.: Information security in the South Australian real estate industry: a study of 40 real estate organisations. Inf. Manag. Comput. Secur. **22**(1), 24–41 (2014). https://doi.org/10.1108/IMCS-10-2012-0060
92. Masrek, M.N.: Assessing information security culture: the case of Malaysia public organization, pp. 1–1 (2018). https://doi.org/10.1109/icitacee.2017.8257663
93. McCracken, G.: Culture and consumption: a theoretical account of the structure and movement of the cultural meaning of consumer goods. J. Consum. Res. **13**(1), 71 (1986). https://doi.org/10.1086/209048
94. McGill, T., Thompson, N.: Old risks, new challenges: exploring differences in security between home computer and mobile device use. Behav. Inf. Technol. **36**(11), 1111–1124 (2017). https://doi.org/10.1080/0144929X.2017.1352028
95. Merchant, K.A., Van der Stede, W.A.: Management control systems: performance measurement, evaluation and incentives. Prentice Hall (2007). https://doi.org/10.5465/ambpp.1964.5067830
96. Merhi, M.I., Ahluwalia, P.: Examining the impact of deterrence factors and norms on resistance to Information Systems Security. Comput. Hum. Behav. **92**, 37–46 (2019). https://doi.org/10.1016/j.chb.2018.10.031

97. Milne, G.R., Labrecque, L.I., Cromer, C.: Toward an understanding of the online consumer's risky behavior and protection practices. J. Consum. Aff. **43**(3), 449–473 (2009). https://doi.org/10.1111/j.1745-6606.2009.01148.x
98. Mylonas, A., Kastania, A., Gritzalis, D.: Delegate the smartphone user? Security awareness in smartphone platforms. Comput. Secur. **34**, 47–66 (2013). https://doi.org/10.1016/j.cose.2012.11.004
99. Ouchi, W.G.: The relationship between organizational structure and organizational control author(s). Adm. Sci. Q. **22**(1), 95–113 (1977)
100. Ouchi, W.G.: A conceptual framework for the design of Organizational Control mechanisms. Manag. Sci. **25**(9), 833–848 (1979). https://doi.org/10.1287/mnsc.25.9.833
101. Parsons, K.M., Young, E., Butavicius, M.A., McCormac, A., Pattinson, M.R., Jerram, C.: The influence of organizational information security culture on information security decision making. J. Cogn. Eng. Decis. Mak. **9**(2), 117–129 (2015). https://doi.org/10.1177/1555343415575152
102. Parsons, K., McCormac, A., Butavicius, M., Pattinson, M., Jerram, C.: Determining employee awareness using the Human Aspects of Information Security Questionnaire (HAIS-Q). Comput. Secur. **42**, 165–176 (2014). https://doi.org/10.1016/j.cose.2013.12.003
103. Pierce, J.L., Kostova, T., Dirks, K.T.: The state of psychological ownership: integrating and extending a century of research. Rev. Gen. Psychol. **7**(1), 84–107 (2003). https://doi.org/10.1037/1089-2680.7.1.84
104. Pierce, J.L., Kostova, T., Dirks, K.T.: Toward a theory of psychological ownership in organizations. Acad. Manag. Rev. **26**(2), 298–310 (2001). https://doi.org/10.5465/AMR.2001.4378028
105. Pierce, J.L., Rubenfeld, S.A., Morgan, S.: Employee ownership: a conceptual model of process and effects. Acad. Manag. Rev. **16**(1), 121 (1991). https://doi.org/10.2307/258609
106. Puhakainen, P., Siponen, M.: Improving employees' compliance through information systems security training: an action research study. MIS Q.: Manag. Inf. Syst. **34**(4), 757–778 (2010). https://doi.org/10.2307/25750704
107. Rhee, H.S., Kim, C., Ryu, Y.U.: Self-efficacy in information security: Its influence on end users' information security practice behavior. Comput. Secur. **28**(8), 816–826 (2009). https://doi.org/10.1016/j.cose.2009.05.008
108. Safa, N.S., Maple, C., Furnell, S., Azad, M.A., Perera, C., Dabbagh, M., Sookhak, M.: Deterrence and prevention-based model to mitigate information security insider threats in organisations. Futur. Gener. Comput. Syst. (2019). https://doi.org/10.1016/j.future.2019.03.024
109. Safa, N.S., Sookhak, M., Von Solms, R., Furnell, S., Ghani, N.A., Herawan, T.: Information security conscious care behaviour formation in organizations. Comput. Secur. **53**, 65–78 (2015). https://doi.org/10.1016/j.cose.2015.05.012
110. Saha, M.: How to See What People Are Doing on Your WiFi (2019). https://techwiser.com/what-others-are-browsing-on-your-wifi/
111. Sahd, L.M., Rudman, R.: Mobile technology risk management. J. Appl. Bus. Res. **32**(4), 1079–1096 (2016). https://doi.org/10.19030/jabr.v32i4.9723
112. Sebescen, N., Vitak, J.: Securing the human: employee security vulnerability risk in organizational settings. J. Am. Soc. Inf. Sci. **68**(9), 2237–2247 (2017). https://doi.org/10.1002/asi
113. Sergueeva, K., Shaw, N.: Improving healthcare with wearables: overcoming the barriers to adoption. In: Lecture Notes in Computer Science (Including Subseries Lecture Notes in Artificial Intelligence and Lecture Notes in Bioinformatics). LNCS, vol. 10293, pp. 209–223 (2017). https://doi.org/10.1007/978-3-319-58481-2_17
114. Sharma, S., Warkentin, M.: Do I really belong? Impact of employment status on information security policy compliance. Comput. Secur. **87**, 101397 (2019). https://doi.org/10.1016/j.cose.2018.09.005
115. Shih, S.P., Liou, J.Y.: Investigate the effects of information security climate and psychological ownership on information security policy compliance. In: Pacific Asia Conference on Information Systems, PACIS 2015—Proceedings (2015)

116. Silic, M., Barlow, J.B., Back, A.: A new perspective on neutralization and deterrence: predicting shadow IT usage. Inf. Manag. **54**(8), 1023–1037 (2017). https://doi.org/10.1016/j.im.2017.02.007
117. Siponen, M.T.: A conceptual foundation for organizational information security awareness. Inf. Manag. Comput. Secur. **8**, 31–41 (2000). https://doi.org/10.1108/09685220010371394
118. Siponen, M.T., Mahmood, M.A., Pahnila, S.: Employees' adherence to information security policies: an exploratory field study. Inf. Manag. **51**(2), 217–224 (2014). https://doi.org/10.1016/j.im.2013.08.006
119. Siponen, M., Vance, A.: Neutralization: new insights into the problem of employee Information Systems security policy violations. MIS Q. **34**(3), 487–502 (2010)
120. Son, J.Y.: Out of fear or desire? Toward a better understanding of employees' motivation to follow IS security policies. Inf. Manag. **48**(7), 296–302 (2011). https://doi.org/10.1016/j.im.2011.07.002
121. Stanton, J.M., Stam, K.R., Mastrangelo, P., Jolton, J.: Analysis of end user security behaviors. Comput. Secur. **24**(2), 124–133 (2005). https://doi.org/10.1016/j.cose.2004.07.001
122. Stanton, J.M., Weiss, E.M.: Electronic monitoring in their own words: an exploratory study of employees' experiences with new types of surveillance. Comput. Hum. Behav. **16**(4), 423–440 (2000). https://doi.org/10.1016/S0747-5632(00)00018-2
123. Stephens, K.K., Ford, J.L.: Unintended consequences of a strategically ambiguous organizational policy selectively restricting mobile device use at work. Mob. Media Commun. **4**(2), 186–204 (2016). https://doi.org/10.1177/2050157915619211
124. Straub, D.W.J., Nance, W.D.: Discovering and disciplining computer abuse in organizations: a field study. Misq **14**, 45–60 (1990). https://doi.org/10.2307/249307
125. Tairov, I.L.: Mobile device management as a component of corporate IT infrastructure. Bus. Manag./Biznes Upravlenie **3**, 60–71 (2019). http://proxy.lib.sfu.ca/login?url=https://search.ebscohost.com/login.aspx?direct=true&db=bth&AN=139040613&site=ehost-live
126. Talib, Y.Y.A.: Intrinsic motivation and Information Systems Security Policy Compliance in Organizations (2015)
127. Thompson, N., McGill, T.J., Wang, X.: "Security begins at home": determinants of home computer and mobile device security behavior. Comput. Secur. **70**, 376–391 (2017). https://doi.org/10.1016/j.cose.2017.07.003
128. Trinkle, B.S., Warkentin, M., Malimage, K., Raddatz, N.: High-risk deviant decisions: does neutralization still play a role? J. Assoc. Inf. Syst. **22**(3), 797–826 (2021). https://doi.org/10.17705/1jais.00680
129. Tsohou, A., Holtkamp, P.: Are users competent to comply with information security policies? An analysis of professional competence models. Inf. Technol. People **31**(5), 1047–1068 (2018). https://doi.org/10.1108/ITP-02-2017-0052
130. Tsohou, A., Karyda, M., Kokolakis, S., Kiountouzis, E.: Managing the introduction of information security awareness programmes in organisations. Eur. J. Inf. Syst. **24**(1), 38–58 (2015). https://doi.org/10.1057/ejis.2013.27
131. Tu, Z., Adkins, J., Zhao, G.Y.: Complying with BYOD security policies: a moderation model based on Protection Motivation Theory. J. Midwest Assoc. Inf. Syst. **2019**(1), 11–28 (2019). https://doi.org/10.17705/3jmwa.000045
132. Tu, Z., Turel, O., Yuan, Y., Archer, N.: Learning to cope with information security risks regarding mobile device loss or theft: an empirical examination. Inf. Manag. **52**(4), 506–517 (2015). https://doi.org/10.1016/j.im.2015.03.002
133. Verkijika, S.F.: Understanding smartphone security behaviors: an extension of the protection motivation theory with anticipated regret. Comput. Secur. **77**, 860–870 (2018). https://doi.org/10.1016/j.cose.2018.03.008
134. Von Solms, B., Von Solms, R.: The 10 deadly sins of information security management. Comput. Secur. **23**(5), 371–376 (2004). https://doi.org/10.1016/j.cose.2004.05.002
135. Vrhovec, S., Markelj, B.: Relating mobile device use and adherence to information security policy with data breach consequences in hospitals. J. Univ. Comput. Sci. **24**(5), 634–645 (2018)

136. Vroom, C., Von Solms, R.: Towards information security behavioural compliance. Comput. Secur. **23**(3), 191–198 (2004). https://doi.org/10.1016/j.cose.2004.01.012
137. Wani, T.A., Mendoza, A., Gray, K.: BYOD in hospitals—security issues and mitigation strategies. In: Australasian Computer Science Week Multiconference Proceedings (ACSW'19) (2019). https://doi.org/10.1145/3290688.3290729
138. Wani, T.A., Mendoza, A., Gray, K.: BYOD usage and security behaviour of hospital clinical staff: an Australian survey. Int. J. Med. Inform. 104839 (2022). https://doi.org/10.1016/j.ijmedinf.2022.104839
139. White, G., Ekin, T., Visinescu, L.: Analysis of protective behavior and security incidents for home computers. J. Comput. Inf. Syst. **57**(4), 353–363 (2017). https://doi.org/10.1080/08874417.2016.1232991
140. Willison, R., Warkentin, M., Johnston, A.C.: Examining employee computer abuse intentions: insights from justice, deterrence and neutralization perspectives. Inf. Syst. J. **28**(2), 266–293 (2018). https://doi.org/10.1111/isj.12129
141. Woon, I.M.Y., Tan, G., Low, R.: A Protection Motivation Theory approach to home wireless security. In: Twenty-Sixth International Conference on Information Systems, vol. 5, pp. 367–380 (2005)
142. Workman, M., Bommer, W.H., Straub, D.: Security lapses and the omission of information security measures: a threat control model and empirical test. Comput. Hum. Behav. **24**(6), 2799–2816 (2008). https://doi.org/10.1016/j.chb.2008.04.005
143. Xue, B., Warkentin, M., Mutchler, L.A., Balozian, P.: Self-efficacy in information security: a replication study. J. Comput. Inf. Syst. (2021). https://doi.org/10.1080/08874417.2021.2015725
144. Xue, B., Xu, F., Luo, X., Warkentin, M.: Ethical leadership and employee information security policy (ISP) violation: exploring dual-mediation paths. Organ. Cybersecur. J.: Pract. Process People **1**(1), 5–23 (2021). https://doi.org/10.1108/ocj-02-2021-0002
145. Yoo, C.W., Rao, H.R.:. Collective security efficacy and group security compliance. In: 35th International Conference on Information Systems "Building a Better World Through Information Systems", ICIS 2014, pp. 1–11 (2014)
146. Zahadat, N., Blessner, P., Blackburn, T., Olson, B.A.: BYOD security engineering: a framework and its analysis. Comput. Secur.Secur. **55**, 81–99 (2015). https://doi.org/10.1016/j.cose.2015.06.011

Chapter 4
Research Methodology

The research methodology is the key component of the study that outlines the detailed activities of the research being undertaken. The research method selected is imperative in setting the right direction for the research and ensures the validity of the data collected. The chapter starts by describing the paradigm and various research for the approach adopted by the current research. The research was carried out by first exploring gaps in the literature. The findings contributed plausible factors to the conceptualization of the research framework. The research then, employed exploratory approach to determine control measures that mitigate BYOD risks qualitatively. These control measures along with the underexplored factors formed the BYOD security policy framework, which was later validated quantitatively in the following phase. The methodology chapter in this thesis describes the research paradigm, research design and the data collection procedures. The sampling technique and size, unit of analysis, research instrument, validity and reliability of the instrument as well as the procedures to analyze the data are also detailed in this chapter. A short summary ends this chapter.

4.1 Research Paradigm

The philosopher, Kuhn [53] developed the term paradigm to characterize a philosophical mode of thought. The research paradigm is viewed as a fundamental set of ideas and assumptions that researchers are ready to accept in order to comprehend and address challenges, it acts as a compass for guiding research activities [35, 56]. It can be characterized by the way scholars respond to three basic questions in terms of ontological, epistemological and methodological questions [35]. Paradigm describes how a researcher views the world, which defines the perspective, thinking, school of thought or set of beliefs that informs the meaning or interpretation of research data [51]. It also provides a conceptual framework or a philosophical one in order to study

© The Author(s), under exclusive license to Springer Nature Switzerland AG 2025
R. Palanisamy et al., *Bring Your Own Device Security Policy Compliance Framework*, Information Systems Engineering and Management 37,
https://doi.org/10.1007/978-3-031-86801-6_4

the world around us [22]. Researchers can ground their inquiries in any number of paradigms and it shapes the theory created for common understanding [3].

The four components of a paradigm are epistemology, ontology, methodology, and axiology [57]. Epistemology is used to describe how knowledge was obtained and how we know if it is the truth. Ontology is about presumptions made in order to believe that something is logical or real. It enables scholars to examine underlying belief system and philosophical assumptions. Methodology refers to the research design, methods, approaches and procedures utilized in the investigation, such as data gathering, participants, instruments and data analysis. When carrying out a research, axiology analyses the ethical problems that must be considered.

There are several research paradigms such as positivism, constructivism and interpretivism, pragmatism among others [17, 52]. A paradigm serves several purposes as listed below and the dimensions for each paradigm are explained in Table 4.1:

- It guides researchers on what they should study, how and when; and how their findings should be interpreted;
- It develops models and theories that permit practitioners solve these issues;

Table 4.1 Dimensions of contrast among three research paradigms [78]

Dimensions of contrast	Qualitative	Mixed method	Quantitative
Methods	Qualitative methods	Mixed methods	Quantitative methods
Paradigm	Constructivism	Pragmatism Transformative	Post positivism Positivism
Research Questions	QUAL research questions	MM research questions	QUAN research questions Research hypotheses
Form of data	Narrative	Narrative and numeric	Numeric
Purpose of research	(often) Exploratory and confirmatory	Exploratory and confirmatory	(often) Confirmatory and exploratory
Role of theory/ logic	Grounded theory Inductive	Both inductive and deductive, inductive and deductive research cycle	Rooted in conceptual framework or theory Hypothetico-deductive model
Typical studies or designs	Ethnographic research designs and others (case study)	MM designs, such as parallel and sequential	Correlational; survey; experimental; quasi-experimental
Sampling	Mostly purposive	Probability, purposive, and mixed	Mostly probability
Data analysis	Thematic strategies; categorical and contextualizing	Integration of thematic and statistical; data conversion	Statistical analysis; descriptive and inferential
Validity	Trustworthiness; credibility; transferability	Inference quality; inference transferability	Internal validity; external validity

- It establishes criteria for tools such as methodology, instruments and data collection that would enable solving these issues;
- It provides the principles, procedures, and methods to be considered when similar issues (phenomena) appear again.

4.1.1 Pragmatism

Philosophers who argued that it was not possible to arrive at the 'truth' about the real world using a single scientific technique, as supported by the Positivist paradigm, or to detect social reality as envisioned by the Interpretivist paradigm, were the initiators of the concept of this paradigm [51]. It uses practical and pluralistic approaches that could allow a combination of methods as a pragmatic way to understand human behavior.

The call for multi method research on the methodological level is one solution to paradigm's incommensurability. Advocates for this combination state that different methods have different strength; thus, the combination would produce more than what each method could offer in isolation [63]. This paradigm promotes relational epistemology where the relationships appropriate for the study is determined by the researcher, a non-singular reality ontology (that there is no single reality and all individuals have their own and unique interpretations of reality), a mixed methods methodology (a combination of quantitative and qualitative research methods) [78].

Pragmatism is not committed to any one system of philosophy and reality [27, 78]. This applies to mixed methods research in that inquirers draw liberally from both quantitative and qualitative assumptions when they engage in their research. Researchers are given the freedom to select the methods, techniques, and procedures of research according to their requirements [18, 48]. In general, a mixed-method approach is a methodology for combining both quantitative and qualitative methods, and they can be executed concurrently or sequentially depending on the purpose of the study [84]. For in-depth understanding and validation, researchers combined components of qualitative and quantitative methodologies [47, 59]. Previous research has examined a handful of published mixed-method studies and uncovered a variety of reasons why researchers employ them [84] such as;

- To improve accuracy of the research data
- Reduce the inherent research bias in a single method approach and boost the research strengths by combining many approaches.
- A way for expanding on the initial findings and developing the analysis utilizing different types of data or methods.

Caracelli and Greene [9] and Greene et al. [34] identified five purposes for conducting mixed method: triangulation, complementary, development, initiation and expansion as detailed in Table 4.2. Triangulation is used when one method to assess a particular phenomenon is deemed to produce a biased and limited results, whereas complementary mixed method study measures the same phenomenon in different aspects. Development uses the results from one method to help develop or inform the other method. Another goal is initiation, which is to uncover paradoxes and new viewpoints, and finally, expansion, which is to broaden the scope and breadth by incorporating various components.

Sandelowski [72] states that mixed method studies are reflected with the combination of techniques which includes the level of sampling, data collection and data analysis. Tashakkori and Teddlie [77, 78] acknowledged the possibility of having both inductive exploratory questions and deductive confirmatory questions in the same study with mixed method research. With inductive reasoning, researchers observe certain phenomena and build a general proposition based on their observation and measures [74]. They are more open ended and follow the reversed process of deductive reasoning. Deduction is a process where researchers arrive at a reasoned conclusion by logical generalization of a known fact. It has a set of techniques for applying testable theories in the real world to evaluate their validity [55]. The first step involves the generation of theories and hypothesis, followed by the operationalization of the concepts so that it can be measured through empirical observation. Next, the researcher identifies the techniques and approaches to measure the constructs

Table 4.2 Purpose and rational to adopt mixed method strategy for evaluation design [34]

Purpose	Rationale
TRIANGULATION seeks convergence, corroboration, correspondence of result from different methods	To increase the validity of constructs and inquiry results by counteracting or maximizing the heterogeneity of irrelevant sources of variance attributable to inherent method bias but also to inquirer bias, bias of substantive theory, biases of inquiry context
COMPLEMENTARY seeks elaboration, enhancement, illustration, clarification of the results from one method with the results from the other method	To increase the interpretability, meaningfulness and validity of constructs and inquiry results by both capitalizing on inherent method strengths and counteracting inherent biases in methods and other sources
DEVELOPMENT seeks to use the results from one method to help develop or inform the other method, where development is broadly construed to include sampling and implementation, as well as measurement decisions	To increase the validity of constructs and inquiry results by capitalizing on inherent method strengths
INITIATION seeks the discovery of paradox and contradiction, new perspectives of frameworks, the recasting of questions or results from one method. with questions or results from the other method	To increase the breadth and depth of inquiry results and interpretations by analyzing them from the different perspectives of different methods and paradigms
EXPANSION seeks to extend the breadth and range of inquiry by using different methods for different inquiry components	To increase the scope of inquiry by selecting the methods most appropriate for multiple inquiry components

(research instrument, sampling plan, data collection methods, analysis and interpretation of the empirical measurements). The final step involves interpreting the meaning of the results [73].

Quantitative research focuses on deductions, confirmation of theory or hypothesis testing, prediction and statistical analysis. Meanwhile, qualitative study focuses on exploring the central or the broad areas of phenomenon and it is a way to explore descriptions and interpretations of people's perspectives or point of view, practices and events that shape them. It allows the researcher flexibility during collection and analysis of narratives, open-ended observations, interviews, focus groups and ethnographies.

Thus, in mixed methods research, qualitative or quantitative has elements of both inductive and deductive components. The researcher may observe patterns and trends in the data that lead them to develop new theories/hypotheses (induction) and deduction provides the agreement about the arguments. Exploratory and confirmatory dichotomies are two types of research. Confirmatory studies are executed when researchers need to test (confirm) a pre-specified relationship or a previously stated hypothesis with statistical inferences [62]. Whereas exploratory studies are utilized when researchers need to define relationships in the most general form before using statistical techniques to estimate their results.

Both exploratory and confirmatory have their strengths and limitations. Though exploratory research can be flexible in developing hypotheses and less requirement for data to support them, it makes it difficult for theory falsification. Confirmatory analysis, on the other hand, delivers information using well-established theories and methodologies, but it is usually guided by preconceived ideas and is unlikely to yield unexpected conclusions.

However, using mixed methods research makes room for the exploratory inductive process that begins with empirical evidence of a particular phenomenon and follows with theorization, generalization and confirmatory deductive process of hypothesis testing [71].

4.1.2 Research Paradigm Selected for This Research

The philosophy of pragmatism, in which the intent is to apply both qualitative and quantitative data collection to explore a problem because the population has been understudied and only little information is available about them. Although BYOD is a global phenomenon, it is currently at an all-time high due to the unprecedented work-from-home requirement due to the pandemic, and is expected to rise further after Covid-19 [86]. The pervasive use, risk, and security policy compliance have not been extensively studied in the existing literature. As a result, this modern phenomenon should be explored using both methodologies because it allows researchers to explore and quantify research problems [41] and it is motivated by the desire to get specific types of insight during the research [85].

Based on the arguments presented in the previous section and from the extant literature, a number of factors can be shown that lead to the choice approach for the current study. Following Creswell [15], this research employed mixed method approach to understand this pervasive phenomenon of security issues and to determine the security policy compliance factors. This was chosen as a single research method would be inadequate to address the four research questions from multiple perspectives. A mixed-methods approach is potentially superior to a single method design, because it can provide rich insights into phenomena and develop novel theoretical explanations.

It can be seen from the existing literature that most of the BYOD security policy compliance studies has a dominant theoretical drive which is positivist in nature [65]. This study chartered the gap by utilizing qualitative approach in the first and second phase of the research as it gave the researcher the freedom to carry out investigation based on her beliefs about the BYOD practice in the public sector, security risk that emanated from it and the factors that mitigated the risk.

As part of the qualitative approach, two systematic reviews were performed to explore and gather factors to form research framework. Additionally, factors that mitigated BYOD security risk were also gathered from the perspective of security experts to be included in the theoretical framework. By determining the factors based on BYOD security risk, it provides rich insights why these strategies are proposed instead of simply adopting previous studies' framework. Then, the researcher used the findings to build the proposed framework to be tested in the following phase of research quantitatively (Fig. 4.1). The findings helped to identify and later measure the significance of the identified variables using existing instrument in the study. As such, mixed method is selected for the purpose of development (as shown in Table 4.2), where it involves sequential use of qualitative and quantitative methods which the first method is used to inform the second [34]. Since, findings from qualitative and quantitative were equally contributed to the core of the research, hence, the study is both inductive and deductive in nature.

Additionally, the study applied confirmatory methods to infer stipulated hypotheses put forth in the study. Statistical measures were used to test the proposed hypotheses, demonstrate the predictive relevance of the BYOD security policy compliance framework. Based on the results, the researcher developed a theoretical framework as the research outcome.

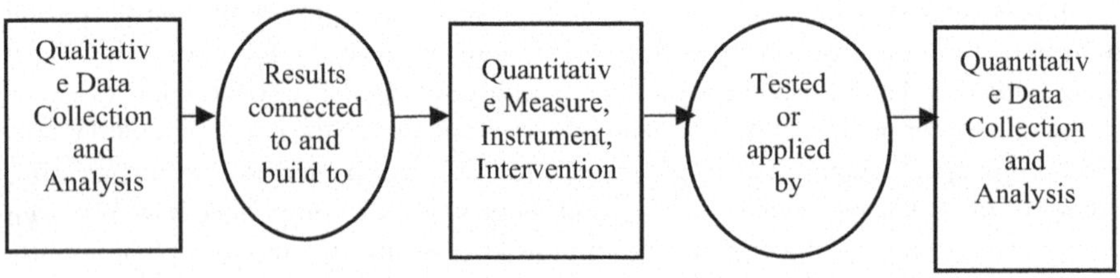

Fig. 4.1 General diagram for exploratory sequential design [15]

Lastly, the current study employs a mixed method approach in order to increase the study's validity and strength while lowering research bias. The approach has been shown to enhance the robustness and accuracy to an overall research design [84]. Findings from the exploratory method contributed to the confirmatory method to develop the BYOD security policy compliance framework (Fig. 4.1). The use of a qualitative method at the beginning is a means to gather constructs to ensure all relevant constructs are studied appropriately [1].

To summarize the research, in order to answer the research questions, the research consists of three phases of studies:

Phase 1: In this phase, the researcher performed a systematic review to identify factors that could influence employees' BYOD security policy compliance.

Phase 2: In the second phase, the researcher performed another systematic review to identify BYOD security risks, and performed semi-structured interviews in order to investigate BYOD practices, security risks, assess the risks to identify critical risks and control measures that mitigate risks. Findings of Phase 2 were mapped into the conceptual research framework to be tested in the following phase.

Phase 3: The conceptual framework was tested using a survey among employees in three CNII agencies. Data obtained was analyzed statistically to determine the factors that influence BYOD security policy compliance intention. These factors form the validated BYOD security policy compliance framework.

4.2 Research Design

According to Bell and Bryman [5], research design provides a framework for data gathering and analysis. The design also reflects considerations regarding the importance of certain aspects of the research process. On the other hand, they consider research methods as the techniques for collecting data which involve specific instruments such as self-completed questionnaires or structured interviews. The purpose of a research design is to ensure that the evidence obtained enables us to answer the initial question as unambiguously as possible. Sekaran and Bougie [74] stated that research design involves a series of rational decision-making choices regarding the purpose of the study (exploratory, descriptive hypothesis testing), its location (the study setting), the type of the investigation, the extent of researcher interference, time horizon and the level to which the data will be analyzed (unit of analysis). Decisions must be taken on the sampling design, data collection methods, and how variables will be assessed and analyzed to test hypotheses (data analysis).

Methods, according to Sekaran and Bougie [74], are crucial component to characterize data gathering. Exploratory sequential design begins with the collection and analysis of qualitative data in the first phase. It explores possible relationships in the most general form and provides insights about directions, strengths and relationship between variables [75, 79], besides allowing freedom for the researcher to gather information pertaining to the study. Building from the exploratory results, the

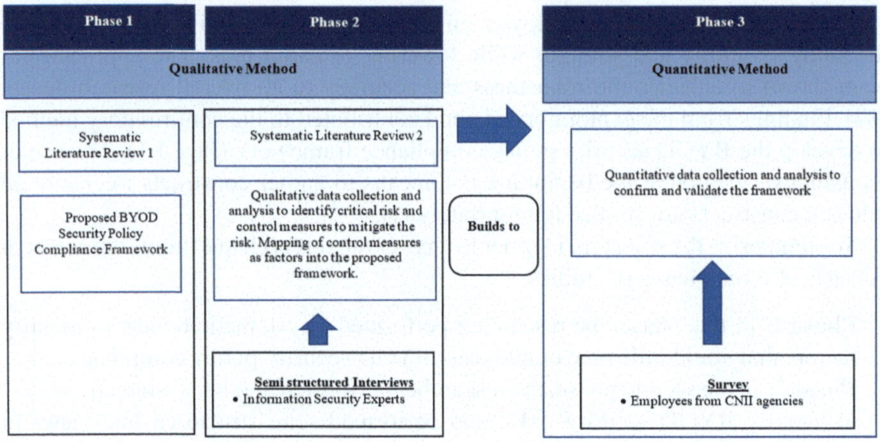

Fig. 4.2 Overview of the research design and methodology

researcher designs and develops the conceptual research framework. The confirmatory research on the other hand is deductive and provides definite answers to research questions.

This study employed an exploratory approach at the beginning by performing systematic reviews to contribute to the conceptual research framework and research hypotheses. Semi-structured interviews and risk assessment with information security experts were conducted to explore the BYOD practices, features, and security risks. The results of the risk assessment reveal risk that pose possible threats to the organization and recommend control measures to mitigate the identified risks. Using the gathered data, the control measures were interpreted as theory constructs in the research framework. The control measures were interpreted as factors that influence BYOD security policy compliance using Organizational Control Theory, Social Cognitive Theory, and security culture.

Subsequently, in the confirmatory phase, hypotheses were validated via quantitative means. Data was collected via online survey from employees from selected CNII agencies and analyzed using both descriptive and confirmatory factor analysis. The analysis confirms the significant factors that predict employees' BYOD security policy compliance behavior. Figure 4.2 depicts the overall research design and method adopted for this research.

4.3 Systematic Literature Review

This study carried out systematic literature review (SLR) following the guidelines proposed by Kitchenham and Charters [50]. This process incorporates all extant research literature that is relevant to answer the research questions. Besides, the review is performed to identify research gaps in the previous literature. The review process involves a number of steps, identifying review questions, setting the inclusion criteria, search strategy, selection of studies, quality assessment, data extraction and data synthesis. As such, systematic literature review was performed to identify BYOD risks and mitigation strategies. Additionally, prominent theories and models were also reviewed extensively to identify BYOD features, security policy compliance and research gaps to contribute to the development of the research model.

4.3.1 SLR Research Questions

a. What are the unique features of BYOD that distinguish BYOD from the traditional computing in organizations?
b. What are the different theoretical lenses that are used in BYOD security policy compliance research?
c. Which BYOD security policy compliance factors are examined in the research?
d. What are the limitations and research gaps within the BYOD security policy compliance research?
e. What are the BYOD security risks caused by security policy noncompliance behavior in organizations?
f. What are the mitigation strategies to influence security policy compliance behavior?

4.3.2 Search Strategy

The review employed both automatic and manual searches to acquire complete results. Manual search was performed after automated search is carried out using research keywords. Search strings were developed based on Kitchenham [49] as a rigorous search process is one main element that distinguishes a systematic review from a traditional review [50]. Table 4.3 shows the keywords and strings developed for the search process. The keywords were searched in the titles and abstract to retrieve as many relevant papers as possible. Six digital databases (Science Direct, ISI Web of Science, IEEE Explorer, ACM Digital Library, Emerald Insight and Scopus) were chosen to identify relevant articles. Manual search using google scholar was also executed to ensure no relevant studies had been missed.

4.3.3 Inclusion and Exclusion Criteria

The inclusion criteria have been set out in the protocol, to ensure that the boundaries of the review questions are clearly defined to ease the process of articles selection. The aim of applying inclusion and exclusion criteria is to confirm that all selected

Table 4.3 Keywords

Keywords and phrases	Synonyms
Bring-your-own-device (BYOD)	Bring your own phone (BYOP), bring your own technology (BYOT), bring your own personal device, IT consumeri*, mobile device*, smartphone*
Policies	BYOD security polic*, information security polic*, bring-your-own-device polic*, information technology polic*, IT polic*
Compliance	Adhere*, complian*, conformity
Non-compliance	Violation, infringement, deterrence
Behavior	Practices, attitude
Security risk	Issues, challenges, threats
Employee	Staff, human resources, user*, employe*

Table 4.4 Inclusion and exclusion

Included articles	• Full text available
	• Published from the year 2009 till 2019
	• Written in English
	• Relevant to the research questions
Excluded articles	• Full text unavailable
	• Non-English
	• Irrelevant to the research questions

studies are important and relevant to the study. The criteria for the search process involved collecting relevant data from journal articles and conference papers which are written in English and have been published in digital databases from the year 2009 to 2019 as presented in Table 4.4.

4.3.4 Study Selection

The next step focused on reading through the titles and abstracts to determine their significance to the research questions. The next step involved skimming the contents of the articles for screening unclear or irrelevant studies. Articles that did not discuss compliance with security policies in a BYOD setting were not considered for inclusion. In a later stage, the snowballing method was utilized to perform a manual search using Google Scholar. This approach was followed to acquire more reliable primary studies.

4.3.5 *Quality Assessment*

The usefulness of the original research's findings and interpretation is determined by assessing the quality of each recognized study using a set of criteria [50]. For this review, a quality assessment was conducted using the questions indicated in Table 4.5 to determine the extent to which a study is appropriate and can provide findings that contribute to the scope of the investigation. The method for determining the article's score was established as follows: Yes = 1, Partial = 0.5, and No = 0 or Unknown (the information is not provided). To begin, a study was given a score of 1 if it totally met the criteria. Second, a value of 0.5 was given to a study when a quality criterion was only partially met by the study in question. Lastly, it received a score of zero if it did not fulfil one of the quality requirements.

4.4 Qualitative Study

There are two components to this phase; (i) data collection and (ii) data analysis. The objective of this phase is to explore BYOD security risks and identify factors that mitigate the security risks in the public sector. The findings were interpreted as plausible factors that could influence employees' BYOD security policy compliance. The qualitative results inform the development of the conceptual research framework that is subsequently tested quantitatively in the following phase.

4.4.1 *Research Ethics*

An application to the University of Malaya Research Ethics Committee (UMREC) was submitted before the study could be carried out. The researcher followed the steps to conform to the UMREC's standards and guidelines to protect human participants from harm. It is also aimed at protecting their privacy and confidentiality of the participants involved in the study. The interview and survey questions were submitted as part of the requirement. After a few months and two rounds of correction, the application was approved. The approval (UM. TNC2/UMREC-758) to conduct research is valid from February 2020 till February 2023 (please refer Appendix A).

The researcher studied the data obtained from each individual and regarded each response as an individual data source. Before beginning the interview, a consent letter was handed to them individually. The purpose of the consent letter was to obtain permission from the participants involved in the interview process, whereby they agree to give full commitment, and to inform them that their rights are protected. Participants who agreed to participate in the interviews and audio recorded, signed the informed consent form, a written agreement before participating in the interview.

Table 4.5 Quality assessment checklist

	Criteria	Description of checklist	Response grading
(QA1)	Does the paper describe features of BYOD?	Yes, more than 2 features are described Partially, 1–2 features are described No, the features are not described	{1, 0.5, 0} (yes, partial, no)
QA2	Is the context addressed in the paper related to BYOD security policy compliance?	Yes, the context is addressed explicitly Partially, the context is addressed broadly in the context of security policy compliance but not explicitly described No, the context is not addressed clearly	{1, 0.5, 0} (yes, partial, no)
QA3	Does the research include a description of analyzed factors involved in security policy compliance	Yes, all constructs are defined and the rationale to use the constructs is provided Partially, at least half of the constructs are defined and the rationale to use the construct is provided No, none of the constructs are defined or the rationale to use them is not provided	{1, 0.5, 0} (yes, partial, no)
QA4	Does the research explore security policy compliance with the help of an established behavioural theory?	Yes, the article explored compliance using a behavioral theory, therefore the theory is clearly defined and the decision to use the theory is explained Partially, the theories are not clearly presented and require additional reading No, the theory used is not defined	{1, 0.5, 0} (yes, partial, no)

All participants were informed that their involvement was voluntary and that they were permitted to exit from the studies at any point of the study.

The researcher also ensured that all subjects were adequately aware of the type of information collected, reason for the survey and information gathering, the manner they were expected to contribute to the study and the benefit of the study.

The survey's participants' personal information will be kept private and secure. The researcher did not reveal the names or personal particulars of any individuals in any of the reports or publication from this study. The fundamental principle of privacy and confidentiality of data was acknowledged for all respondents. Additionally, the researcher ensured that the acquired data, including interview transcripts, field notes, and audio recordings, are maintained in a safe location that is only accessible by the researcher. After five years from the completion of the study, the data will be destroyed and shredded.

4.4.2 Data Collection

Qualitative method allows us to gain deep understanding of the underlying BYOD practices and security risks among employees. This study collected qualitative data through interviews with subject matter experts. In order to investigate the prevalent risks that are present in workplaces where BYOD policies are widely implemented, data was gathered by conducting interviews with industry experts.

The first step was to select a representative sample of study participants. A purposive sample was selected as the participants were information security experts with extensive knowledge and experience managing the organization's security. Purposive sampling is the selection of participants who meet the criteria that are outlined by the researcher [8, 66]. Gentles et al. [33] proposed that purposive sampling was the deliberate selection of appropriate sources for data collection to fulfil the research aim. Security experts can easily contribute to a deeper understanding of the identification of BYOD critical security concerns and factors that minimize the risks.

The sample size was determined by the acquired information and the saturation level. Data saturation includes the adequacy and accuracy of the obtained data from the participants [64]. Data saturation is the stage where the data obtained results in no new themes, insights, perspectives, or information in the analysis [43]. There were no agreed significance level or specific number of interviews that needs to be carried out in a qualitative research [61]. As such, 8 security experts were involved in this study. A sufficient sample size permits the researcher to classify the findings into comparable patterns and themes [64]. The experts were coded E1 to E8 throughout the research process to protect the confidentiality of the participants' personal information. Four experts were chosen from three different security agencies that are accountable for managing information security for the Malaysian Public Sector. Additionally, three experts were chosen from a security organization that manages national cyber security operations and one expert who is responsible of managing in-house security in an agency. Although the exact job titles of the various information security experts varied, all of those who were interviewed described themselves as information security practitioners. The people who were interviewed ranged in age from their late 30s to their early 50s and included both males and females. The identities of the experts have been concealed in order to foster open discussions. Researcher faced challenges to gather participants for this study as some of the experts were reluctant to discuss BYOD practices and security risks in their organizations. In order to lessen the pressure of and protecting interviewees, the identities of the organizations are not revealed as described in Table 4.6.

Table 4.6 Purposive homogenous sampling of security experts

	Identifier	Organization name	Informant
1	E1	Security Office	Security Manager (male)
2	E2, E5	Cyber Security Agency	Risk and Compliance Manager (male), Security Manager (female)
3	E3	Central Agency	Security Consultant (male)
4	E4	Ministry	IT Security Compliance Manager (female)
5	E6, E7, E8	Cyber Security Consultation Agency	IT Helpdesk Head (female), Security Researcher (male), Security Compliance Manager (female)

4.4.3 Interviews

Interview is a useful technique since informants are able to describe and explain in great detail the processes that they go through on a daily basis [54]. Furthermore, the value of this strategy lies in the fact that it enables us to unearth additional data in order to gain a deeper comprehension of information security issues from the point of view of information security managers who are subject matter experts (SME).

The semi-structured interview method was employed because it encourages in-depth inquiry, permits additional conversations from each participant's perspective [30] to explore participants' thoughts, feelings, and generates new knowledge [54]. The open-ended questions allowed the researcher to engage in dialogue with participants to acquire detailed responses [87, 88]. The interview guide was developed after performing an in-depth examination of the existing research on the topic, and it was evaluated by two information security experts from the university (please refer interview guide in Appendix D). The experts determined the validity and clarity of the questions by providing researcher with feedback to improve and revise the interview questions to eliminate potential confusion and biases, eliminating redundant questions and using the right word choices.

The interviews took place between December 2019 and February 2020. The length of time for each interview ranged between 30 and 40 min. These allowed ample time for the participants to answer the questions in depth. All interviews were voice-recorded with the participants' consent. The interviews were conducted in English and Malay.

Credibility is the truth of the data obtained from the participants and enhanced by the researcher's interpretation by describing experiences and verifying the findings [13, 42]. One of the ways to demonstrate credibility is by applying the purposive sampling technique for the participants' selection who fulfilled the criteria of contributing necessary information for the research. In addition, the researcher facilitated eliciting open replies from the participants by conducting one-on-one interviews in an environment where each participant felt at ease to share their perspectives [82]. Prolonged engagement is a process of building trust and rapport with informants to enable them to willingly share their information [13]. The researcher also actively listened to each participant to deepen understanding of the responses that addressed the phenomenon for achieving credibility.

As the responses became more saturated over time, no additional interviews were required [14]. Data saturation was attained when researcher has gathered all of the information necessary and no additional interviews yield any new information [31].

4.4.4 Risk Assessment

The following step is to obtain an assessment of the security risks from the viewpoints of security experts. During this phase, a list of security risks is provided to the experts in order to provide them with a more comprehensive view of security risks from existing literature. The risk list is a result from a literature review paper that categorized security risks posed by employees' BYOD security policy non-compliance behavior into the People, Process and Technology (PPT) model. Risk assessment is conducted to measure and analyze the risk level in order to safeguard the operations and services of the organizations. Performing risk assessment calculates severity of attacks by assessing their likelihood of a risk occurring and estimating the potential damage from successful attacks on personal devices. The security experts evaluated risks impact and likelihood to rank the severity of the risks. The risk assessment form was adapted from the Malaysia Public Sector Information Security Risk Assessment Guideline which clearly details the methods and techniques to be deployed in implementing a systematic and effective information security risk assessment [60]. Each expert was given a risk assessment form to evaluate the impact and likelihood. This task took the experts approximately 20 min to accomplish. After determining the likelihood and impact of a particular risk, the risk rating table is examined to determine the risk level. The qualitative research process to achieve research objective two (RO2) (Table 4.7) is summarized in Fig. 4.3.

Table 4.7 Qualitative data analysis plan

Research objectives	Research questions	Method	Scale	Analysis
(RO2) To identify the risks caused by employees' non-compliance behavior towards the BYOD security policy and control measures to mitigate the risks	RQ2: what are the risks caused by employees' non-compliance behavior towards the BYOD security policy and what are the control measures to mitigate the risk?	Interviews Risks assessment	Semi-structured questions	Manual coding to form thematic analysis

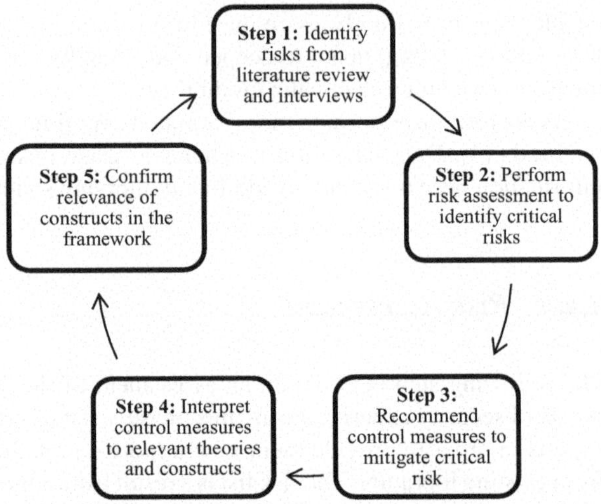

Fig. 4.3 Qualitative research process

4.4.5 Qualitative Data Analysis

A qualitative data analysis of the data obtained during the interviews were used to provide interpretative resources, insights, and explanation of the study and to develop research model that will be tested quantitatively.

The researcher transcribed each participant's interview responses verbatim, and performed member checking to ensure that the final version of the transcript was agreed upon by the researcher and interviewees. Transcription reviews allowed the participants the opportunity to examine and, if necessary, revise the researcher's interpretations of their responses. Confirmability is achieved by having respondents validate the transcripts. Participants are able to review their responses and correct them if necessary [36] and this helps to clarify misunderstandings [4]. All transcriptions of interviews were emailed to the respondents.

Exploratory analysis was used in the research, and its processes include reading through all of the transcripts, identifying and underlining the concepts that constitute emerging themes, and then drawing conclusions from those themes [16]. Multiple readings were performed on the data to uncover phrases, words and highlighting crucial parts that reveal the thoughts and perceptions of the experts in accordance with the stated research objectives (see Table 4.9). To determine the themes and subthemes, the data was coded sentence by sentence [14]. The thematic analysis is a fundamental technique for qualitative analysis that searches interview data for themes or patterns [7].

Transferability is described as readers directly involved in the qualitative study are able to associate with other individuals' settings or group experiences. The findings from the interviews are not generalized facts but rather understandings of how the security managers who were interviewed interpreted the risks associated with BYOD and the severity of those risks. It is possible that these findings can be applied to other public sector organizations all over the world. The research processes are explained in sufficient detail for researchers from different disciplines to comprehend and reproduce the study [42].

As for the risk assessment, the number of risks ranked 'Low', Medium, and 'High' were tabulated in a Excel worksheet and ranked accordingly. Each risk rated as high by five or more (a majority), out of a total of eight experts, was classified as a critical risk in this risk assessment process. For each critical risk, experts recommended control measures to mitigate the risk.

The findings of the interviews and the risk assessments are discussed in greater depth in Chap. 5.

4.5 Quantitative Study

While qualitative data are limited when it comes to statistical analysis and conforming hypotheses, findings derived from qualitative studies can be used to support and build existing theories. Hence, quantitative study is required to validate the findings from previous phase. Following the completion of the research survey, the results are analyzed to discover significant factors. The quantitative process is depicted in Fig. 4.4.

Fig. 4.4 Process flow to validate the research framework

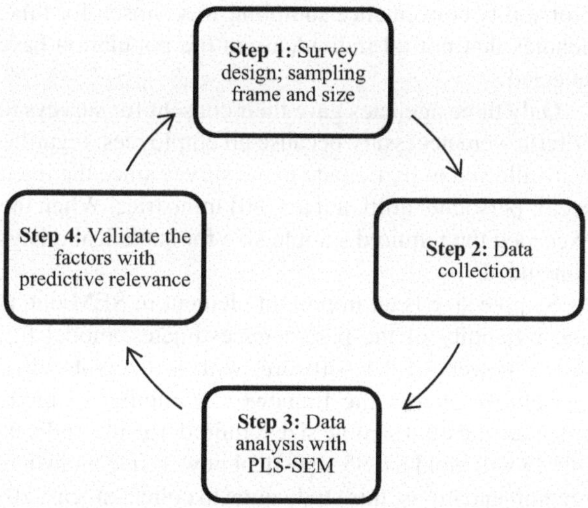

4.5.1 *Data Collection*

The data was collected from three CNII agencies from the health, defense and finance sectors in Putrajaya, Malaysia from September 2020 till February 2021. The online survey was hosted on Survey Monkey platform and distributed through the official email of the government organization to its employees. Prior to data collection, a number of considerations must be made, as detailed below;

4.5.1.1 Population

This study's population consists of employees working in the CNII agencies in Malaysia. There are nine critical sectors as listed in the NACSA website. However, only three agencies from three CNII sectors (health, defence and finance) gave permission and agreed to broadcast the survey to their employees.

4.5.1.2 Sampling Techniques

Quantitative studies require adequate and the right sample or respondents for empirical analysis [74]. During the pandemic, data gathering in the public sector was extremely difficult, making probability sampling impractical. Therefore, non-probability convenience sampling was chosen for this study. This type of sampling denotes that not all individuals in the population have an equal opportunity to be selected.

Only three agencies gave their consent for surveys to be conducted. No exclusion criteria were necessary because all employees, regardless of scheme, level, or grade, were allowed to participate in the survey since the intended respondents were public sector personnel affiliated to CNII ministries. When the total number of respondents exceeded the required sample size for statistical analysis, the data collection works concluded.

Sample size is an important element in SEM due to their effects on the validity and reliability of the parameter estimates, model fit and statistical power. Using the G*Power 3.1.9.4 software, with a moderate effect size of 0.15, and an error probability of 5%, the framework's number of predictors were used as input to calculate the total sample size required. Results indicate that a minimum sample size of 129 will yield a 0.95 statistical power. In conclusion, considering the sample size recommendations, this study aims to collect at least 200 responses from the targeted demographic because the heterogeneity of the community necessitates a sufficient sample size for further data analysis.

4.5.1.3 Procedures

There are few processes used during data collection for this study. Prior to carrying out the final survey, the processes were meticulously planned. This section will discuss the sequence of approaches used to acquire and analyze quantitative data.

4.5.1.4 Expert Reviews

Pretesting is a test to assess the appropriateness and understandability of the questions that will be included in the real study, using a limited number of respondents. The purpose of the pretest was to obtain feedback relating to the validity of the survey instrument from experienced academic and industry experts [18]. It also aims to avoid poor data quality and to reduce or prevent the removal of items during the measurement model evaluation [2].

Seven specialists in information security and research methodology pretested the survey. The specialists included two IS security experts from International Islamic University Malaysia (IIUM), two IS experts from University Utara Malaysia (UUM), University Teknologi MARA, and a research method expert from University Sains Malaysia. Two additional industry specialists came from the government, specifically Malaysian Administrative Modernisation and Management Planning Unit (MAMPU) and National Institute of Public Administration (INTAN) (please refer Appendix E). All the experts were asked to comment on a variety of factors, including the suitability of the employed scale, the clarity or ambiguity of construct definitions, and the representation of items.

Feedback from the experts resulted in the following changes; some of the statements measuring each construct were rephrased to improve clarity and relevancy. Redundant items were removed. Some ambiguous words were replaced and lengthy sentences were shortened.

4.5.1.5 Questionnaire Design

The research instrument was created by adapting existing validated measurements. The items were adjusted to ensure that respondents understood them and that the researcher could determine the appropriate response. The phrasing and length of the items were altered to ensure that they represent the essence of the issue.

a. **Content and purpose of questions**: As the purpose of this study is to measure the factors impacting security policy compliance, the items in the key constructs for this research were chosen based on the findings from the interviews and a thorough examination of the literature.
b. **Wording and Language**: Careful attention was paid to ensure that the wording of the items can be understood by the respondents. Using the guideline by Podsakoff et al. [68], the questionnaire was carefully designed in terms of the word and

language use. It is important to ensure that the questions are simple, easy to read and straight to the point. The language used should also be not too complex. In addition, it is necessary to avoid ambiguity, uncertainty, vagueness, and the use of double-barreled questions by ensuring that each question focuses on a single subject.

c. Furthermore, every question on the questionnaire was adapted from surveys performed in English-speaking western nations, hence, the questions were translated into Malay language. The questionnaire was designed in both English and Malay language to ensure that respondents understood the statements.
d. **Type and Form of questions**: Several questions are used in the measurement items. Close ended questions were used in questions on the independent variables, demographic profile and the dependent variable. The use of close ended questions will help respondents to make informed, yet quick decisions by selecting the most relevant option. Besides that, all of the items in the questionnaire are positive questions and no negative questions. As suggested by Podsakoff et al. [68], the researcher must eliminate vagueness by providing clear examples, simplifying complex and compound questions and use simple language words to ensure the questions are easily understood by the respondents.
e. **Common Method Variance**: CMV is a possible concern even though a few procedural remedies have been taken before distributing the questions. Therefore, to minimize the CMV, two techniques were used to test the impact of CMV towards collected data by using Harman's single factor test, full collinearity to test CMV.

4.5.1.6 Scale Development

As explained, all constructs were adapted from established questionnaires and no single item was used in the model. The constructs were selected based on the suitability of the questions to suit the research context as informed by the systematic reviews. The questions adapted were previously validated and empirically tested by the prior studies. Table 4.8 summarizes the constructs and the number of items used to measure each construct. All constructs were operationalized using Likert scales. A Likert type scale is commonly used to measure different latent constructs. All of the items on the independent variables used a seven-point Likert scale except for Top Management, Device Monitoring and SETA which used a five-point Likert scale. Meanwhile, the dependent variable, which is the security policy behavior used Yes/No response.

Table 4.8 Constructs and number of items for pilot test

No.	Constructs	No. of items	Scale
1	Top management commitment	3	Five point
2	Device monitoring	3	Five point
3	Policy mandatoriness	4	Seven-point
4	Specification of policy	4	Seven-point
5	Evaluation of policy	4	Seven-point
6	Self-efficacy	3	Seven-point
7	BYOD IT support	4	Seven-point
8	BYOD SETA	8	Five-point
9	Psychological ownership	7	Seven-point
10	BYOD security policy compliance behavior	5	(yes/no)
	Total	45	

Table 4.9 Items to measure specification of policy, evaluation of policy and policy mandatoriness

Indicator	Measurement statement	Adapted from
PM1	*I am required to secure my personal devices according to the organization's ICT Security Policy and procedures*	[6]
PM2	*It is expected that I will play an active role in securing my devices from cyber-attacks (hacking, virus infection, data corruption, etc.)*	
PM3	*There is an understanding that I will comply with organization's ICT Security Policy and procedures*	
PM4	*Regulatory compliance requirements (ISMS (ISO27001)) emphasizes the need for me to follow the organization's ICT Security Policy to the best of my ability*	
SOSP1	*I am familiar with the organization's ICT Security Policy, procedures and guidelines*	
SOSP2	*I am required to know a lot of existing written procedures and general practices to secure my personal devices*	
SOSP3	*There are written rules regarding ICT Security Policy and procedures at my organization*	
SOSP4	*The organization's existing ICT Security Policy and guidelines cover how to protect my personal devices*	
EVSP1	*Managers in my department frequently evaluate my security behaviors*	
EVSP2	*Managers regularly examine data relating to how well I follow security policies and procedures (system log)*	
EVSP3	*Managers formally evaluate me and my colleagues regarding compliance with ICT security policy*	
EVSP4	*Managers assess whether I follow organizational information security procedures and guidelines*	

4.5.2 *Operationalization and Measurement of Constructs Deployed for the Pilot Test*

The accurate measurement scales were identified and chosen for each operational variable as described below.

4.5.2.1 Organizational Control Theory (Specification of Policy, Evaluation of Policy, and Policy Mandatoriness)

Specification of policy is described as a security policy that has well-defined objectives to ensure all employees abide by the set of rules and procedures [24]. Evaluation of policy verifies if the expected result has been attained or whether the employee has demonstrated the specified behaviors according to the security policies [32]. Policy Mandatoriness denotes the degree to which individuals perceive that compliance with existing security policies and procedures is compulsory or expected by organizational management [6]. The items as detailed in Table 4.9.

4.5.2.2 Social Cognitive Theory (BYOD IT Support and Self-efficacy)

The BYOD IT support rendered by the organization is deemed to improve employees' ability and their confidence to manage a specific task [32] to resolve device issues efficiently [46]. The items are listed in Table 4.10.

Table 4.10 Items to measure BYOD IT support and self-efficacy

Indicator	Measurement statement	Adapted from
SE1	*It is easy for me to comply with my organization's ICT Security Policy and apply all security measures*	[81]
SE2	*I have the capability to comply with my organization's ICT Security Policy to protect the confidential organization data stored in my own devices*	
SE3	*I can apply all required security measures and controls from my personal devices to prevent unauthorized access to my organization's network and internal systems*	
ITS1	*My organization has an IT team that provides support to employees who use their personal devices for work related tasks*	[46]
ITS2	*An IT support team is available to assist employees with their personal devices when needed*	
ITS3	*The IT support team helps employees in following the requirements as stated in the ICT Security Policy*	
ITS4	*The IT support team helps employees if they encounter security problems on their personal devices*	

Table 4.11 Items to measure security culture

Indicator	Measurement Statement	Adapted from
TM1	*Senior management actively champions information security goals*	[19]
TM2	*Top management considers information security as an important organizational priority*	
TM3	*Top managers adhere to ICT Security Policy themselves*	
DMON1	*I believe that my organization monitors any modification or altering of computerize organizational data by employees*	
DMON2	*I believe that my organization reviews log of employees' personal device computing activities on a regular basis*	
DMON3	*I believe that employee personal devices computing activities are monitored by my organization*	

4.5.2.3 Security Culture (Top Management and Device Monitoring)

Security culture is a critical component to direct employees, contract staff, and vendors to reduce risks on information assets [21]. In this study, security culture is defined by top management and device monitoring. The six (6) items measuring security culture are listed in Table 4.11.

4.5.2.4 BYOD SETA

The BYOD SETA reflects on information security training equips employees with confidence and important skills that help them to execute responsibilities as specified in the ISP [20]. There are eight (8) items adapted from [46], as listed in Table 4.12.

4.5.2.5 Psychological Ownership

Psychological ownership is a feeling of having a possessive connection and feelings of responsibility with targets that individuals feel are theirs [25, 67]. The seven (7) items are shown in Table 4.13 as follow.

4.5.2.6 Security Policy Compliance Behavior

Information security compliance activities are a series of activities that comply with managerial and technical information security policies as shown in Table 4.14.

Table 4.12 Items to measure BYOD SETA

Indicator	Measurement statement	Adapted from
SETA1	*My organization provides training to improve employees' awareness of information security issues regarding the use of personal devices for work related tasks*	[46]
SETA2	*In my organization, employees are briefed on the risk to the organization of not securing mobile devices with a proper password*	
SETA3	*In my organization, employees are briefed on the risk to the organization of not installing recommended antivirus software on their mobile devices*	
SETA4	*In my organization, employees are briefed on the risk to the organization of installing banned applications on their mobile devices*	
SETA5	*In my organization, employees are briefed on the personal risk of not securing personal devices with a proper password*	
SETA6	*In my organization, employees are briefed on the personal risk of not installing recommended antivirus software on their personal devices*	
SETA7	*In my organization, employees are briefed on the personal risk of installing banned applications on their personal devices*	
SETA8	*My organization explains to employees about responsibilities in using personal devices for work related tasks*	

Table 4.13 Items to measure psychological ownership

Indicator	Measurement Statement	Adapted from
PSYO1	*I feel a high degree of ownership for my personal devices and its contents*	[80]
PSYO2	*The information stored in my device is very important to me*	
PSYO3	*I personally invested a lot in my personal devices (e.g., time, effort, money)*	
PSYO4	*I personally invested a lot in the software/applications on my personal devices (e.g., time, effort, money)*	
PSYO5	*When I think about it, I see an extension of my life in my device*	
PSYO6	*I have personalized my device to better suit the way I use it*	
PSYO7	*I see my device as an extension of myself*	

4.5.3 Pilot Test

Pilot test was a small-scale preliminary study conducted by researchers to evaluate survey prior to its distribution to the targeted respondents (Kent 2001). It is considered a prerequisite for validating the survey measurements, and is conducted with a small group of respondents [73]. Weaknesses in the survey design, for instance ambiguous scale item or questions that does not match respondents work context can be identified

Table 4.14 Items to measure security behavior

Indicator	Measurement statement	Adapted from
SECB1	*I have recent backups of my personal devices*	[80]
SECB2	*I have installed security software (e.g., anti-virus/anti malware) on my personal devices*	
SECB3	*I have enabled automatic updating of my personal devices' software*	
SECB4	*I use security software (e.g., anti-virus/anti malware) for my personal devices*	
SECB5	*My personal devices are secured by a password*	

from this process. Pilot test evaluates the respondents' ability to understand and answer the survey without requiring further clarification or assistance.

In 2020, pilot test was performed in which the questionnaire was randomly sent to 100 respondents from various ministries who were excluded from the population sample. They were invited to participate in the pilot test via email and WhatsApp™ which contained the hyperlink to the online survey. Some of the respondents were contacted for a brief interview to discuss the possible shortcomings of the survey instrument and ways to improve it. The respondents were queried about the clarity and suitability of the questions. Five of the respondents mentioned about the difficulty to answer both the constructs; SETA and Security Behavior questions which was too detailed and they were unable to answer them. Analysis of the pilot test also found many missing values for those constructs as 25 respondents left the SETA section blank.

Since many of the agencies do not have comprehensive BYOD clauses in their ICT security policy, these items were difficult for them to answer. The items of the security behavior do not match security policy contents as each agency uses general clauses for BYOD. These items were meant to measure security behavior and not security policy compliance. Hence, the items were revised to "Intention to Comply" which is adapted from [28]. With the early findings from the pilot test, survey items for both constructs were revised and shown to the respondents to finalize the survey questions (Tables 4.15, 4.16, and 4.17) (please refer Appendix F).

4.5.4 Final Survey

It took researchers around two months to create and validate the final survey. The Deputy Secretary General of each agency granted approval for the survey to be distributed to all personnel. The data collection process took five months, from November to March. The pandemic exacerbated the situation and posed a significant challenge to the data collection process because most staffs were remote-working. It was challenging that the target respondents were from vital agencies, and they were

Table 4.15 Amendments to BYOD SETA for the final survey items

Pilot test items [46]	*Amended final survey items* [45]
SETA1: My organization provides training to improve employees' awareness of information security issues regarding the use of personal devices for work related tasks	***SETA1****: My organization periodically briefs on the issues of information security breaches from the use of personal devices for work related tasks through awareness campaigns (e-mail, brochure/seminar/workshops)*
SETA2: In my organization, employees are briefed on the risk to the organization of not securing mobile devices with a proper password	***SETA2****: My organization keeps me updated on information security breaches and preventive measures regarding the use of personal devices for work related tasks*
SETA3: In my organization, employees are briefed on the risk to the organization of not installing recommended antivirus software on their mobile devices	***SETA3:*** *My organization keeps educating employees on their personal devices' security responsibilities*
SETA4: In my organization, employees are briefed on the risk to the organization of installing banned applications on their mobile devices	***SETA4:*** *My organization's information security training programs keep me well-informed against security threats from the use of personal devices for work related tasks*
SETA5: In my organization, employees are briefed on the personal risk of not securing personal devices with a proper password	***SETA5:*** *My organization's information security awareness programs develop necessary skills to adopt protective behavior*
SETA6: In my organization, employees are briefed on the personal risk of not installing recommended antivirus software on their personal devices	
SETA7: In my organization, employees are briefed on the personal risk of installing banned applications on their personal devices	
SETA8: My organization explains to employees about responsibilities in using personal devices for work related tasks	

Table 4.16 Amendments to the security policy compliance scale items

Pilot test items [80]	*Amended Final survey items* [28]
SECB2: I have recent backups of my device	***INT2****: I intend to protect my personal devices used for work according to the requirements of my organization's BYOD security policy*
SECB3: I have enabled automatic updating of my computer software	***INT3****: I intend to carry out my responsibilities prescribed in the BYOD security policy of my organization when I use my personal devices for work purposes*
SECB4: I use security software (anti-virus/anti malware)	
SECB5: My device is secured by a password	

Table 4.17 Number of items for final survey

Construct	Code	Items	Likert scale	Source
Specification of policies	SOSP1	I am familiar with the organization's IT security policies, procedures, and guidelines	7-point	[6]
	SOSP2	I am required to know a lot of existing written procedures and general practices to secure my personal devices		
	SOSP3	There are written rules regarding security policies and procedures at the organization		
	SOSP4	The organization's existing policies and guidelines cover how to protect my personal devices		
Evaluation of security policy compliance	EVSP1	Managers in my department frequently evaluate my security behaviors	7-point	[6]
	EVSP2	Managers regularly examine data relating to how well I follow security policies and procedures		
	EVSP3	Managers formally evaluate me and my colleagues regarding compliance with security policies		
	EVSP4	Managers assess whether I follow organizational security procedures and guidelines		
Perceived mandatoriness	PM1	I am required to secure my system according to the organization's documented policies and procedures	7-point	[6]
	PM2	It is expected that I will take an active role in securing my personal devices from cyber-attacks (hacking, virus infection, data corruption, etc.)		
	PM3	There is an understanding that I will comply with organization security policies and procedures		
	PM4	Regulatory compliance requirements (ISMS:ISO 9001:2013) emphasize the need for me to follow the organization's IT security policies, procedures and guidelines to the best of my ability		
Self-efficacy	SE1	It is easy for me to comply with my organization's ICT Security Policy and apply all security measures	7-point	[81]
	SE2	I have the capability to comply with my organization's ICT Security Policy to protect the confidential organization data stored in my own devices		

(continued)

Table 4.17 (continued)

Construct	Code	Items	Likert scale	Source
	SE3	I can apply all required security measures and controls from my personal devices to prevent unauthorized access to my organization's network and internal systems		
BYOD IT support	ITS1	My organization has an IT team that provides support to employees who use their personal devices for work related tasks	7-point	[46]
	ITS2	An IT support team is available to assist employees with their personal devices when needed		
	ITS3	The IT support team helps employees in following the requirements as stated in the ICT Security Policy		
	ITS4	The IT support team helps employees if they encounter security problems on their personal devices		
SETA	SETA1	My organization periodically briefs on the issues of information security breaches from the use of personal devices for work related tasks through awareness campaigns (e-mail, brochure/ seminar/ workshops)	7-point	[45]
	SETA2	My organization keeps me updated on information security breaches and preventive measures regarding the use of personal devices for work related tasks		
	SETA3	My organization keeps educating employees on their personal devices' security responsibilities		
	SETA4	My organization's information security training programs keep me well-informed against security threats from the use of personal devices for work related tasks		
	SETA5	My organization's information security awareness programs develop necessary skills to adopt protective behaviour		
Psychological ownership	PSYO1	I feel a high degree of ownership for my device and its contents	7-point	[80]
	PSYO2	The information stored in my device is very important to me		
	PSYO3	I personally invested a lot in my device (e.g., time, effort, money)		

(continued)

Table 4.17 (continued)

Construct	Code	Items	Likert scale	Source
	PSYO4	I personally invested a lot in the software/ applications on my device (e.g., time, effort, money)		
	PSYO5	When I think about it, I see an extension of my life in my device		
	PSYO6	I have personalized my device to better suit the way I use it		
	PSYO7	I see my device as an extension of myself		
Top management commitment	TM1	Senior management actively champions security goals	5-point	[19]
	TM2	Top management considers information security an important organizational priority		
	TM3	Top managers adhere to security policies themselves		
Device monitoring	MON1	I believe that my organization monitors any modification or altering of computerized data by employees	5-point	[19]
	MON2	I believe that my organization reviews log of employees' computing activities on a regular basis		
	MON3	I believe that employee computing activities are monitored by my organization		
Intention to comply	INT1	I intend to comply with the requirements of the BYOD security policy of my organization	7-point	[28]
	INT2	I intend to protect my personal devices used for work according to the requirements of my organization's BYOD security policy		
	INT3	I intend to carry out my responsibilities prescribed in the BYOD security policy of my organization when I use my personal devices for work purposes		
Total items	40			

tasked with a multitude of ad hoc duties in managing the COVID-19 impact on the country. In addition, it was hard to obtain the population framework for each ministry. Agencies that granted permission sent survey invitation emails to all workers who were primarily working from home at the time of data collection. The email includes a link to a SurveyMonkey-hosted survey. The researcher conducted follow-ups via email reminder, telephone calls and WhatsApp messages (Table 4.18).

Table 4.18 Quantitative data analysis plan

Research objectives	Research questions	Measurement	Scale	Analysis
RO3: To examine the relationships between the factors and employees' compliance behavior towards BYOD security policy	What is the relationship between the identified factors and employees' compliance behavior towards BYOD security policy?	Questionnaire	**Likert scale (1–7)** 1. Strongly disagree 2. Disagree 3. Slightly disagree 4. Neutral 5. Slightly agree 6. Agree 7. Strongly agree **Likert scale (1–5)** 1. Strongly disagree 2. Disagree 3. Neutral 4. Agree 5. Strongly agree	PLS-SEM analysis
RO4: To validate the BYOD Security Policy Compliance Framework	How does the framework predict employee's compliance intention towards the BYOD security policy?			PLS-SEM analysis

4.5.5 Data Analysis

To establish a general understanding of the data, SPSS was used to analyze demographic information, and frequency and percentage analyses were performed on the data gathered. In addition, SPSS was also used to perform Harman's single factor test to detect common method bias test. SmartPLS version 3.0 [70] was used in the second phase of data analysis in order to test the measurement model and structural model of this study. SmartPLS is a variance-based software, which is suitable for causal predictor analysis, especially in the condition of high and low theoretical information. PLS-SEM permits the simultaneous testing of the measurement model (relationships between constructs and indicators) and the structural model (theoretical relationship among constructs) with an adequate PLS measurement model, wherein the hypotheses may be tested by evaluating the PLS structural model. The explanatory power of the PLS structural model based on the amount of variance endogenous constructs in the research model. PLS-SEM does not impose normality requirement on the data. SEM methodology is useful in behavioral and social sciences. Applying

Table 4.19 Measurement model analysis guidelines

No.	Assessment	Name of index	Guidelines
1	Internal consistency	Composite reliability (CR)	CR > 0.9 (not desirable) CR > 0.7–0.9 (satisfactory) CR > 0.6 (for exploratory research)
2	Indicator reliability/ factor loadings	Indicator loadings	Loading 0.708 or higher is recommended, but loadings > 0.7, 0.6, 0.5 or 0.4 is adequate if other items have higher loadings to complement
3	Convergent validity	Average variance extracted (AVE)	AVE > 0.50
4	Discriminant validity	HTMT criterion	HTMT 0.85 HTMT inference using bootstrapping technique [44]: does 90% bootstrap confidence interval of HTMT include the value − 1 < HTMT < 1

SEM to test the hypothesized relationships between factors allows a complete investigation of all the hypothesized relationships simultaneously. The current study is more about predictive analysis whereby the conceptual model of the study can be categorized as prediction-oriented modelling. For that reason, PLS-SEM was applied to test the hypotheses.

The first phase of data analysis deals with the data screening procedures in order to ensure that the data has been correctly entered, and to test the normality of the data. The second phase was the application of a two-stage PLS-SEM process. Stage one was to assess the measurement properties of PLS-SEM, which involved the assessment of each latent variable and the test of reliability and the validity of the measurement properties. The second stage involved specification of the path relationship between the underlying theoretical latent constructs. If all the measurements are valid, the structural model can be used for hypotheses testing. SmartPLS 3.0 was used to conduct a PLS-SEM analysis to identify the significant predictors of compliance behavior in order to achieve the third study objective (RO3). Details are provided in Tables 4.19 and 4.20.

4.5.5.1 Justification of Using PLS-SEM

According to the guideline listed in Table 4.19, this study chose the PLS-SEM for the research model assessment. The technique was chosen due to the following conditions below:

- The current study incorporates the OCT, SCT, and Security Culture with the addition of new variables to construct the research model; thus, the focus of this

Table 4.20 Summary of indices for structural model assessment [69]

No.	Assessment	Name of index	Guidelines	Literature support
1	Lateral collinearity	Variance inflator factor (VIF)	VIF ≤ 3.3 VIF ≤ 5.0	[23] [38]
2	Path coefficient	Path coefficient	p value < 0.01 t value > 2.58 (two-tailed) t value > 2.33 (one-tailed) p value < 0.05 t value > 1.96 (two-tailed) t value > 1.645 (one-tailed) p value < 0.10 t value > 1.645 (two-tailed) t value > 1.28 (one-tailed)	
3	R^2	Coefficient of determination	0.26—substantial 0.13—moderate 0.02—weak 0.67—substantial 0.33—moderate 0.19—weak	[12] [10]
4	f^2	Effect size to R^2	0.35—substantial 0.15—moderate 0.02—weak	[12]
5	Q^2	Stone Geisser Q^2 predictive relevance	> 0 indicates that exogenous constructs have predictive relevance towards the endogenous constructs	

research is on theory development and prediction, which adhered to the PLS-SEM rules of thumb.

- PLS is suitable to be applied to complex structural models with a large number of constructs.

In most studies on BYOD security policies, PLS-SEM has been used extensively as the data analysis technique. All constructs in the research framework except security culture are reflective constructs. Security culture was conceptualized as reflective-formative higher order construct. Reflective constructs aim to identify measures that are inter-correlated and have strong internal consistency. The following subsections discuss the guidelines used to access the measurement model. The software employs a two-stage procedure: (i) measurement model assessment; and (ii) structural model estimation. These stages are further elaborated below:

4.5.5.2 Measurement Model Assessment

Item reliability and validity are the focus of Stage 1 (Measurement Model Assessment). According to Hair et al. (2017b), researchers may proceed with structural model testing after all the required criteria are met by the measurement models. Measurement items are deemed reliable if they consistently measure the concepts they are intended to measure. On the other hand, measurement items are deemed to be valid if they measure their intended concept well. Convergent validity and discriminant validity were performed to test whether the measurement model in this study is reliable and valid [58]. Figure 4.5 list out the step-by-step procedures involved in the measurement model.

Indicator Reliability (Outer Loadings)

According to [83], if an indicator truly measures what it is intended to measure, then the indicators are deemed reliable. Indicator reliability is conducted to test the extent to which the indicators are consistent with what they intend to measure [83]. This study applied the cut off value of 0.5 to determine the significance value of each indicator [37]. Loading values equal to and greater than 0.5 are acceptable, if the summation of loadings in high loading scores, contributing to AVE scores of greater than 0.5.

Fig. 4.5 Steps to perform measurement model

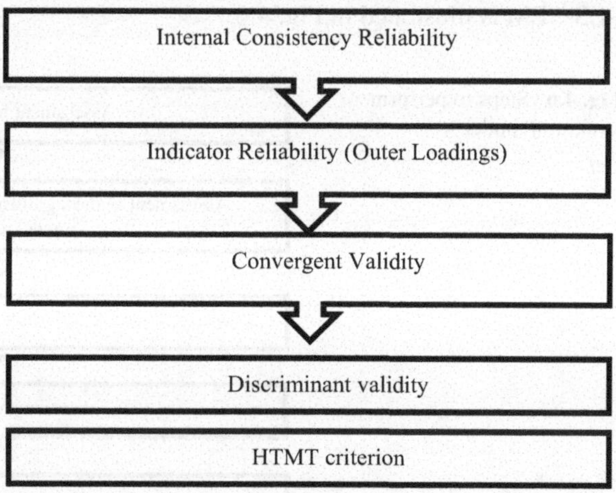

Convergent Validity

Convergent validity can normally be determined using the Average Variance Extracted (AVE). Urbach and Ahlemann [83] stated that convergent validity takes two measures that are supposed to be measuring the same construct and show that they are related.

Discriminant Validity

Discriminant validity shows that two measures that are not supposed to be related are in fact, unrelated [83]. Once the component score of each LV correlates with that of all other items, cross-loading is present [10]. There is no interchangeability between indicators of different constructs, if compared to any other construct, the loading of each indicator for its designated construct is higher. Henseler et al. [44] recommended testing the HTMT value, whether it is significantly different from 1.00 via bootstrapping. HTMT could also be tested at a lower threshold value such as 0.85 or 0.90, depending on the study context [29]. A summary of the guidelines for assessing measurement model validity is provided in Table 4.19.

4.5.5.3 Structural Model Estimation

The structural model was assessed after the establishment of the measurement model. The model's capabilities to predict one or more target constructs were determined in this step [38]. The five (5) steps performed for assessing the structural model in PLS-SEM is illustrated in Fig. 4.6.

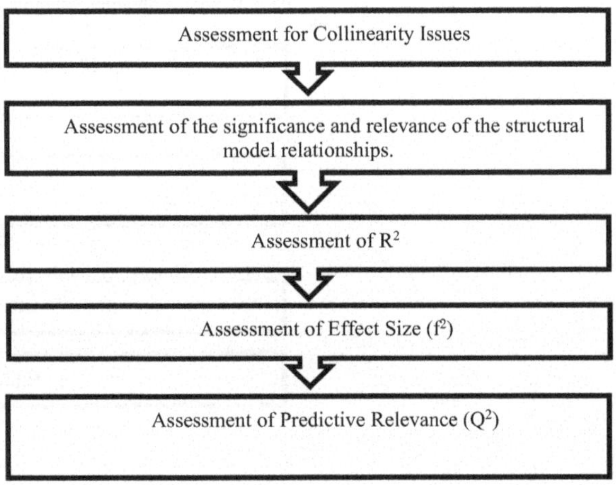

Fig. 4.6 Steps to perform structural model

Assessment of Collinearity Issues

To ensure that the data introduces no bias to the regression results, the structural relationship were assessed and collinearity issues determined. Studies have stated that predictor constructs could have a high probability of collinearity issues if they have a VIF value of above 5; whereas other say that lower VIF values (3–5) could also indicate collinearity problems. Therefore, the researcher should aim for a VIF value of 3 and lower. One method to address collinearity is to create models that are theory-supported and with a higher order [38].

Structural Model Relationship

As part of the hypothesis testing in the study, the results of each path relationship in the framework were generated using a bootstrapping procedure (a nonparametric test in PLS). In bootstrapping, the original sample is replaced with repeated random sampling to obtain the standard errors for hypothesis testing and to create a bootstrapped sample [39]. Chin [11] suggested performing bootstrapping with 1000 re-samples. To test significance level, the bootstrapping function in SmartPLS 3.0 software was used to generate the *t*-statistics for all paths. The bootstrapping was set to 0.05 significance level, with a one tailed test and 5000 subsamples. Critical values for the one tailed test are 2.33 at the 1% significance level ($\alpha = 0.01$), 1.645 at the 5% significance level ($\alpha = 0.05$) and 1.28 at the 10% significance level ($\alpha = 0.1$) [69].

Coefficient of Determination (R^2)

Next, the framework's predictive accuracy was evaluated using the coefficient of determination (R^2). A framework's explanatory power can be determined from its R^2 value. The coefficient of determination (R^2) is calculated through the squared correlation between the exogenous and endogenous variables. This coefficient measures the variance in the endogenous constructs that can be explained by each of the exogenous variables that are linked to it. The predictive power of the model could also be determined from the R^2 value. In other words, the coefficient of determination (R^2) ranging from 0 to 1, shows the predictive power of the model. Hair et al. [38] mentioned that the model has a predictive accuracy if the R^2 values are higher, though it is challenging to provide the 'Rule of Thumb' for an acceptable R^2 values depending on the research discipline and model complexity. Chin [10] suggest R^2 value of 0.67 indicating significant predictive power, 0.333 average predictive power and 0.19 as weak predictive power. Falk and Miller [26] recommended that R^2 be at a minimum of 0.10 for an adequate explanatory power of the model indicating a less stringent requirement.

Effect Size (f^2)

The f^2 measures the relative impact of a predictor construct on endogenous constructs. The calculation of f^2 should be assessed if the deleted exogenous variable has a strong influence on the endogenous variable in addition to the R^2 values [38]. According to Sullivan and Feinn [76], apart from the *p*-value, it is also important to report effect size (substantive significance) and *p*-value (statistical significance). Cohen [12] guideline is used to measure effect size. Table 4.20 shows the effect size as defined by Cohen [12] where the values of 0.02, 0.15, and 0.35 correspond to small, medium and large effect sizes, respectively.

Predictive Relevance Assessment (Q^2)

Finally, the blindfolding procedure, as suggested by Hair, et al. (2017b), was used to assess the predictive relevance (Q^2) of the model. A sample reuse technique is used in blindfolding process. Hair et al. [40] stated that, "PLS-SEM exhibits predictive relevance, it can accurately predict the data points of indicators in reflective measurement models of the endogenous constructs and endogenous single item constructs". If $Q^2 > 0$, then, the model has sufficient predictive relevance.

4.5.6 Chapter Summary

This research employed both qualitative and quantitative approach to understand the factors that influence employees' BYOD security policy compliance. In this study, systematic reviews and qualitative study findings were used as a basis to develop the conceptual framework. Systematic reviews were conducted to identify underexplored factors that may contribute to security policy compliance. Additionally, in the qualitative study, BYOD security risks were assessed to identify critical risks and then identify control measures to mitigate the risks. These findings were used to develop the conceptual research framework. Hypotheses were developed based on the relationships between the identified variables and were verified using quantitative approaches.

Description of data collection procedure and method were included comprising of the identified population, sampling procedure as well as the empirical testing of the research hypotheses using various statistical techniques. The research instrument was ensured to have validity and reliability by running a pre-test. The chapter ends with the requirements for the PLS-SEM statistical analysis.

References

1. Al-Qeisi, K.I.: Analyzing the Use of UTAUT Model in Explaining an Online Behaviour: Internet Banking Adoption (2009)
2. Ali Memon, M., Ting, H., Ramayah, T., Chuah, F., Cheah, J.-H.: A Review of the methodological misconceptions and guidelines related to the application of structural equation modeling: a Malaysian scenario. J. Appl. Struct. Equ. Model. **1**(1), 1–8 (2017)
3. Babbie, E.: The Practice of Social Research. Wadsworth Publishing Company (1995)
4. Baškarada, S.: Qualitative case study guidelines. Qual. Rep. (2014). https://doi.org/10.46743/2160-3715/2014.1008
5. Bell, E., Bryman, A.: The ethics of management research: an exploratory content analysis. Br. J. Manag. **18**(1), 63–77 (2007). https://doi.org/10.1111/j.1467-8551.2006.00487.x
6. Boss, S.R., Kirsch, L.J., Angermeier, I., Shingler, R.A., Boss, R.W.: If someone is watching, I'll do what I'm asked: mandatoriness, control, and information security. Eur. J. Inf. Syst. **18**(2), 151–164 (2009). https://doi.org/10.1057/ejis.2009.8
7. Braun, V., Clarke, V.: Using thematic analysis in psychology. Qual. Res. Psychol. **3**, 77–101 (2006)
8. Campbell, S.: Conducting case study research. Am. Soc. Clin. Lab. Sci. **28**(3), 201–205 (2015). https://doi.org/10.29074/ascls.28.3.201
9. Caracelli, V.J., Greene, J.C.: Crafting mixed-method evaluation designs. N. Dir. Eval. **1997**(74), 19–32 (1997). https://doi.org/10.1002/ev.1069
10. Chin, W.W.: The partial least squares approach for structural equation modeling. Mod. Methods Bus. Res. **1998**, 295–336 (1998)
11. Chin, W.W.: How to write up and report PLS analyses. In: Handbook of Partial Least Squares (2010). https://doi.org/10.1007/978-3-540-32827-8
12. Cohen, J.: Statistical Power Analysis for the Behavioural Sciences. Lawrence Erlbaum (1988)
13. Cope, D.G.: Methods and meanings: credibility and trustworthiness of qualitative research. Oncol. Nurs. Forum **41**(1), 89–91 (2014). https://doi.org/10.7748/nr2009.07.16.4.40.c7160
14. Corbin, J., Strauss, A.: Basics of Qualitative Research: Techniques and Procedures for Developing Grounded Theory, vol. 31, issue 4. Sage Publications (2008). https://doi.org/10.1177/1350507600314007
15. Creswell, J.: A Concise Introduction to Mixed Methods Research. Sage Publications (2015)
16. Creswell, J.: Educational Research: Planning, Conducting, and Evaluating Quantitative and Qualitative Research. Pearson (2014)
17. Creswell, J., Plano-Clark, V.L., Gutmann, M.L., Hanson, W.E.: Advanced mixed methods research designs. In: Tashakkori, A., Teddlie, C. (eds.) Handbook of Mixed Methods in Social and Behavioral Research, pp. 209–240 (2003). http://www.sagepub.com/upm-data/19291_Chapter_7.pdf
18. Creswell, J.W., Creswell, J.D.: Research Design: Qualitative, Quantitative and Mixed Method Approaches. Sage Publications (2018)
19. D'Arcy, J., Greene, G.: The multifaceted nature of security culture and its influence on end user behavior. In: IFIP TC 8 International Workshop on Information Systems Security Research, pp. 145–157 (2009)
20. D'Arcy, J., Hovav, A., Galletta, D.: User awareness of security countermeasures and its impact on information systems misuse: a deterrence approach. Inf. Syst. Res. **20**(1), 79–98 (2009). https://doi.org/10.1287/isre.1070.0160
21. da Veiga, A., Martins, N.: Defining and identifying dominant information security cultures and subcultures. Comput. Secur. **70**, 72–94 (2017). https://doi.org/10.1016/j.cose.2017.05.002
22. Deshpande, R.: "Paradigms lost": on theory and method in research in marketing. J. Mark. **47**(4), 101–110 (1983). https://doi.org/10.2307/1251403
23. Diamantopoulos, A., Siguaw, J.A.: Formative versus reflective indicators in organizational measure development: a comparison and empirical illustration. Br. J. Manag. **17**(4), 263–282 (2006). https://doi.org/10.1111/j.1467-8551.2006.00500.x

24. Diesch, R., Pfaff, M., Krcmar, H.: A comprehensive model of information security factors for decision-makers. Comput. Secur. **92** (2020). https://doi.org/10.1016/j.cose.2020.101747
25. Dyne, L.V., Pierce, J.L.: Psychological ownership and feelings of possession: three field studies predicting employee attitudes and organizational citizenship behavior. J. Organ. Behav. **25**(4), 439–459 (2004). http://www.jstor.org/stable/4093721
26. Falk, F.R., Miller, N.B.: A Primer for Soft Modeling. University of Akron Press (1992)
27. Feilzer, M.Y.: Doing mixed methods research pragmatically: implications for the rediscovery of pragmatism as a research paradigm. J. Mixed Methods Res. **4**(1), 6–16 (2010). https://doi.org/10.1177/1558689809349691
28. Feng, G., Zhu, J., Wang, N., Liang, H.: How paternalistic leadership influences IT security policy compliance: the mediating role of the social bond. J. Assoc. Inf. Syst. **20**, 1650–1691 (2019). https://doi.org/10.17705/1jais.00581
29. Franke, G., Sarstedt, M.: Heuristics versus statistics in discriminant validity testing: a comparison of four procedures. Internet Res. **29**(3), 430–447 (2019). https://doi.org/10.1108/IntR-12-2017-0515
30. Furneaux, B., Wade, M.: An exploration of organizational level information systems discontinuance intentions. MIS Q.: Manag. Inf. Syst. **35**(3), 573–598 (2011). https://doi.org/10.2307/23042797
31. Fusch, P.I., Ness, L.R.: Are we there yet? Data saturation in qualitative research. Qual. Rep. **20**(9), 1408–1416 (2015). https://doi.org/10.46743/2160-3715/2015.2281
32. Galvez, S.M., Shackman, J.D., Guzman, I.R., Ho, S.M.: Factors affecting individual information security practices. SIGMIS-CPR'15 **26**(8), 555–557 (2015). https://doi.org/10.1145/2751957.2751966
33. Gentles, S.J., Charles, C., Ploeg, J., McKibbon, A.K.: Sampling in qualitative research: insights from an overview of the methods literature. Qual. Rep. **20**(11), 1772–1789 (2015). https://nsuworks.nova.edu/tqr/vol20/iss11/5%0Ahttps://cpb-us-e1.wpmucdn.com/sites.nova.edu/dist/a/4/files/2016/01/Gentles_Sampling-2016-01-16.ppt.pdf
34. Greene, J.C., Caracelli, V.J., Graham, W.F.: Toward a conceptual framework for mixed-method evaluation designs. Educ. Eval. Policy Anal. **11**(3), 255–274 (1989). https://doi.org/10.3102/01623737011003255
35. Guba, E.: The alternative paradigm dialog. In: Guba, E. (ed.) The Paradigm Dialog, pp. 17–30. Sage (1990)
36. Hagens, V., Dobrow, M.J., Chafe, R.: Interviewee Transcript Review: assessing the impact on qualitative research. BMC Med. Res. Methodol. **9**(1), 1–8 (2009). https://doi.org/10.1186/1471-2288-9-47
37. Hair, J.F., Black, W.C., Babin, B.J., Anderson, R.E.: Multivariate Data Analysis. Cengage (2018). https://doi.org/10.1198/tech.2007.s455
38. Hair, J.F., Hult, T.M., Ringle, C.M., Sarstedt, M.: A Primer on Partial Least Squares Structural Equation Modeling (PLS-SEM). Sage Publications (2017). https://doi.org/10.1080/1743727x.2015.1005806
39. Hair, J.F., Ringle, C.M., Sarstedt, M.: PLS-SEM: indeed a silver bullet. J. Mark. Theory Pract. **19**(2), 139–152 (2011). https://doi.org/10.2753/MTP1069-6679190202
40. Hair, J.J.F., Sarstedt, M., Hopkins, L., Kuppelwieser, V.G.: Partial least squares structural equation modeling (PLS-SEM): an emerging tool in business research. Eur. Bus. Rev. **26**(2), 106–121 (2014). https://doi.org/10.1108/EBR-10-2013-0128
41. Hall, B., Howard, K.: A synergistic approach: conducting mixed methods research with typological and systemic design considerations. J. Mixed Methods Res. **2**(3), 248–269 (2008). https://doi.org/10.1177/1558689808314622
42. Hammarberg, K., Kirkman, M., De Lacey, S.: Qualitative research methods: when to use them and how to judge them. Hum. Reprod. **31**(3), 498–501 (2016). https://doi.org/10.1093/humrep/dev334
43. Harsh, S.: Purposeful sampling in qualitative research synthesis. Qual. Res. J. **11**(2), 63–75 (2011). https://search.proquest.com/docview/920894910/41BC710066204835PQ/1?accountid=13042

44. Henseler, J., Ringle, C.M., Sarstedt, M.: A new criterion for assessing discriminant validity in variance-based structural equation modeling. J. Acad. Mark. Sci. **43**(1), 115–135 (2015). https://doi.org/10.1007/s11747-014-0403-8
45. Hina, S., Selvam, D.D.D.P., Lowry, P.B.: Institutional governance and protection motivation: theoretical insights into shaping employees' security compliance behavior in higher education institutions in the developing world. Comput. Secur. **87**, 101594 (2019). https://doi.org/10.1016/j.cose.2019.101594
46. Hovav, A., Putri, F.F.: This is my device! Why should I follow your rules? Employees' compliance with BYOD security policy. Pervasive Mob. Comput. **32**, 35–49 (2016). https://doi.org/10.1016/j.pmcj.2016.06.007
47. Johnson, R.B., Onwuegbuzie, A.J.: Toward a definition of mixed methods research. J. Mixed Methods Res. **1**(2), 112–133 (2007). https://doi.org/10.1177/1558689806298224
48. Jokonya, O.: The Significance of mixed methods research in information systems research the significance of mixed methods research. In: Proceedings of the Eleventh Midwest Association for Information Systems Conference, Milwaukee, Wisconsin, 19–20 May, pp. 1–5 (2016). http://aisel.aisnet.org/mwais2016/20
49. Kitchenham, B.: Procedures for Performing Systematic Reviews. In Joint Technical Report (2004). https://doi.org/10.5144/0256-4947.2017.79
50. Kitchenham, B., Charters, S.: Guidelines for Performing Systematic Literature Reviews in Software Engineering (2007). https://doi.org/10.1145/1134285.1134500
51. Kivunja, C., Kuyini, A.B.: Understanding and applying research paradigms in educational contexts. Int. J. High. Educ. **6**(5), 26 (2017). https://doi.org/10.5430/ijhe.v6n5p26
52. Klein, H.K., Myers, M.D.: A set of principles for conducting and evaluating interpretive field studies in information systems. MIS Q. **23**(1), 67 (1999). https://doi.org/10.2307/249410
53. Kuhn, T.S.: The structure of scientific revolutions. In: Neurath, O. (ed.) International Encyclopedia of Unified Science, vol. 2, issue 2 (1970). https://doi.org/10.1046/j.1440-1614.2002.t01-5-01102a.x
54. Kvale, S.: InterViews: An Introduction to Qualitative Research Interviewing. Sage Publications (1996)
55. Lancaster, G.: Research methods in management: a concise introduction to research in management and business consultancy (2005)
56. Lincoln, Y.S., Guba, E.G.: Paradigmatic controversies, contradictions, and emerging confluences. In: Denzin, N.K., Lincoln, Y.S. (eds.) The Sage Handbook of Qualitative Research, vol. 4, issue 2, pp. 163–188 (2005)
57. Lincoln, Y.S., Guba, E.G.: Naturalistic Inquiry. Sage Publications (1985)
58. Lowry, P.B., Gaskin, J.: Partial least squares (PLS) structural equation modeling (SEM) for building and testing behavioral causal theory: when to choose it and how to use it. IEEE Trans. Prof. Commun. **57**(2), 123–146 (2014). https://doi.org/10.1109/TPC.2014.2312452
59. Mackenzie, N., Knipe, S.: Research dilemmas: paradigms, methods and methodology. Issues Educ. Res. **16**(2), 193–205 (2006)
60. MAMPU: Cyber security and disaster response and recovery (2010). https://www.malaysia.gov.my/portal/content/30526
61. Marshall, B., Cardon, P., Poddar, A., Fontenot, R.: Does sample size matter in qualitative research? A review of qualitative interviews in is research. J. Comput. Inf. Syst. **54**(1), 11–22 (2013). https://doi.org/10.1080/08874417.2013.11645667
62. Meyers, L.S., Gamst, G., Guarino, A.J.: Applied Multivariate Research: Design and Interpretation. Sage Publications (2005). https://proxy.library.tamu.edu/login?url=https://search.ebscohost.com/login.aspx?direct=true&db=cat03318a&AN=tamug.4019139&site=eds-live
63. Morgan, D.L.: Practical strategies for combining qualitative and quantitative methods: applications to health research. Qual. Health Res. **8**(3), 362–376 (1998). https://doi.org/10.1177/104973239800800307

64. Morse, A., McEvoy, C.: Qualitative research in sport management: case study as a methodological approach. Qual. Rep. **19**(1990), 1–13 (2014). https://doi.org/10.46743/2160-3715/2014.1032
65. Palanisamy, R., Norman, A.A., Kiah, M.L.M.: Compliance with Bring Your Own Device security policies in organizations: a systematic literature review. Comput. Secur. 101998 (2020). https://doi.org/10.1016/j.cose.2020.101998
66. Patton, M.Q.: Qualitative Research & Evaluation Methods, 4th edn. Sage Publications (2015)
67. Pierce, J.L., Kostova, T., Dirks, K.T.: The state of psychological ownership: integrating and extending a century of research. Rev. Gen. Psychol. **7**(1), 84–107 (2003). https://doi.org/10.1037/1089-2680.7.1.84
68. Podsakoff, P.M., MacKenzie, S.B., Lee, J.Y., Podsakoff, N.P.: Common Method Biases in Behavioral research: a critical review of the literature and recommended remedies. J. Appl. Psychol. **88**(5), 879–903 (2003). https://doi.org/10.1037/0021-9010.88.5.879
69. Ramayah, T., Cheah, J.-H., Chuah, F., Ting, H., Memon, M.A.: Partial Least Squares Structural Equation Modeling (PLS-SEM) using SmartPLS 3.0. Handbook of Market Research, 2nd edn. Pearson Malaysia Sdn Bhd. (2018)
70. Ringle, C.M., Wende, S., Becker, J.M.: SmartPLS 3.0 (2015). www.smartpls.com
71. Rocco, T.S., Bliss, L.A., Gallagher, S., Perez-Prado, A.: Taking the next step: mixed methods research in organizational systems. Inf. Technol. Learn. Perform. J. **21**(1), 19–29 (2003). http://citeseerx.ist.psu.edu/viewdoc/download?doi=10.1.1.122.1050&rep=rep1&type=pdf
72. Sandelowski, M.: Focus on research methods: combining qualitative and quantitative sampling, data collection, and analysis techniques in mixed-method studies. Res. Nurs. Health **23**(3), 246–255 (2000). https://doi.org/10.1002/1098-240x(200006)23:3%3c246::aid-nur9%3e3.0.co;2-h
73. Sekaran, U.: Research Methods for Business: A Skill Building Approach. Wiley (2003)
74. Sekaran, U., Bougie, R.: Research Methods for Business: A Skill Building Approach, 7th edn. Wiley (2016). https://doi.org/10.1007/978-94-007-0753-5_102084
75. Stebbins, R.A.: Exploratory Research in the Social Sciences: What Is Exploration? Sage Publications (2001)
76. Sullivan, G.M., Feinn, R.: Using effect size—or why the P value is not enough. J. Grad. Med. Educ. **4**(3), 279–282 (2012). https://doi.org/10.4300/jgme-d-12-00156.1
77. Tashakkori, A., Teddlie, C.: Issues and dilemmas in teaching research methods courses in social and behavioural sciences: US perspective. Int. J. Soc. Res. Methodol.: Theory Pract. **6**(1), 61–77 (2003). https://doi.org/10.1080/13645570305055
78. Teddlie, C., Tashakkori, A.: Foundations of Mixed Methods Research: Integrating Quantitative and Qualitative Approaches in the Social and Behavioral Sciences. Sage (2009)
79. Thompson, B.: Exploratory and confirmatory factor analysis: understanding concepts and applications. Appl. Psychol. Meas. **31**(3), 245–248 (2007). https://doi.org/10.1177/0146621606290168
80. Thompson, N., McGill, T.J., Wang, X.: "Security begins at home": determinants of home computer and mobile device security behavior. Comput. Secur. **70**, 376–391 (2017). https://doi.org/10.1016/j.cose.2017.07.003
81. Tu, Z., Adkins, J., Zhao, G.Y.: Complying with BYOD security policies: a moderation model based on Protection Motivation Theory. J. Midwest Assoc. Inf. Syst. **2019**(1), 11–28 (2019). https://doi.org/10.17705/3jmwa.000045
82. Turner, D.W.: Qualitative interview design: a practical guide for novice investigators. Qual. Rep. **15**(3), 754–760 (2010). https://doi.org/10.46743/2160-3715/2010.1178
83. Urbach, N., Ahlemann, F.: Structural equation modeling in information systems research using partial least squares. J. Inf. Technol. Theory Appl. **11**(2), 5–40 (2010)
84. Venkatesh, V., Brown, S.A., Bala, H.: Bridging the qualitative-quantitative divide: guidelines for conducting mixed methods research in information systems. Manag. Inf. Syst. Q. **37**(1), 21–54 (2013). https://doi.org/10.25300/MISQ/2013/37.1.02
85. Weber: Editor's Comments: The Rhetoric of Positivism versus Interpretivism: A Personal View. MIS Q. **28**(1), iii (2004). https://doi.org/10.2307/25148621

86. Wood, L.: Global BYOD and Enterprise Mobility Market Trajectory & Analytics Report 2021: New World of Opportunities Ahead for Enterprise Mobility Strategies (2021). ResearchAndMarkets.Com. https://www.businesswire.com/news/home/20210310005531/en/Global-BYOD-and-Enterprise-Mobility-Market-Trajectory-Analytics-Report-2021-New-World-of-Opportunities-Ahead-for-Enterprise-Mobility-Strategies---ResearchAndMarkets.com
87. Yin, R.K.: Qualitative Research from Start to Finish, 2nd edn. (2016)
88. Yin, R.K.: Case Study Research and Applications: Design and Methods, 6th edn. Sage Publications (2018)

Chapter 5
Results and Discussion

Previous chapter explained the justification for conducting qualitative research and described the data collection methods involved, followed by a description of the qualitative data analysis process, which covers the selection of the participants and the interviews that were conducted. This chapter first presents the findings of the qualitative research in which the data was analyzed manually. Data collected from the interviews were subjected to thematic analysis and each of the research findings was interpreted into relevant theories and concepts, followed by discussion of the qualitative findings. The findings contribute to the conceptual research framework that will be tested quantitatively in the following phase of study.

This chapter then describes the quantitative analysis conducted and presents the empirical findings used to test the research hypotheses. The quantitative phase consists of data editing, coding and screening, prior to the PLS-SEM analysis. Following a general overview of the survey respondents, the results of Confirmatory Factor Analysis (CFA), which assesses reliability and construct validity, are presented. The results of the structural model to test the hypotheses developed is reported in this chapter. In the following section, the predictive relevance of the research model is discussed. This chapter finishes with a discussion on the study's findings.

5.1 Qualitative Research Findings

Qualitative research was performed to identify plausible factors influencing BYOD security policy compliance. Firstly, through the interviews with the security experts, the real scenario of employees' BYOD practice and the security risks it posed were understood. Risk assessment was conducted based on the extensive list of BYOD security risk to evaluate risk severity and experts identified factors that mitigate the

© The Author(s), under exclusive license to Springer Nature Switzerland AG 2025

R. Palanisamy et al., *Bring Your Own Device Security Policy Compliance Framework*,
Information Systems Engineering and Management 37,
https://doi.org/10.1007/978-3-031-86801-6_5

critical risks. These factors were later interpreted based on relevant theories to be tested in the next phase of research with quantitative, confirmatory study.

With manual coding, thematic analysis was performed to organize and analyze the data from the interviews. All data collected from the interviews were transcribed manually. The transcribed interview data were then deconstructed and categorized to determine important concepts. The main and sub-themes were from the content of the interviews developed through the coding process.

Each theme's theoretical contribution was researched and further refined until a satisfactory set of themes were identified. The purpose of undertaking this approach was to derive thematic areas by converging the codes into sub-themes and main themes. The frequency and percentage of the data that fit each theme was calculated and presented in Table 5.1.

5.1.1 BYOD Security Risks

The ever-evolving BYOD trend has created new threats, vulnerabilities, and gaps in organizations. Based on the outcomes of the interviews, the following risks have been grouped into three thematic areas: People, Process, and Technology (PPT) Model.

5.1.1.1 People Risks

a. **Information sharing behavior**

The information sharing behavior was a recurring theme in all of the interviews. The majority of employees share organizational data from their personal devices with an unauthorized or third-party app, which poses a risk to the organization's data security. *E1* stated that *"with mobile device which is equipped with communication applications, for example WhatsApp, this becomes a huge problem when employees have a chatgroup, and they share classified documents, some even snap a picture and share. It becomes worse when some individuals in the group blast that particular information to other groups. This becomes the cause for data leakage. In some cases, the information is modified and circulated around"*. According to *E1*, restructuring plans for ministries and departments were revealed over WhatsApp following the country's thirteenth election. This event occurred as a result of someone taking a screenshot of a confidential document and sharing it via WhatsApp. Using applications such as WhatsApp, users can send text messages, video, audio communications, and photos to personal or professional contacts instantaneously and simultaneously through the Internet. The second occurrence, according to *E2*, was the leak of the prime minister's budget speech before its formal release. It is rather disturbing to see that some employees mishandle confidential information, such as tender documents sharing them via various information sharing platforms. *"These employees just snap photos of confidential documents and share" (E4, E6)*. *E2* also indicated that with

Table 5.1 Expert identity and percentage of risk themes elicited

Thematic area	Main themes	Sub-theme	Theme description	Respondents	%
BYOD security risks	People	Information sharing behavior	Employees are able to easily exchange information with other in online community	E1, E2, E4, E5, E6	63
		Complacent behavior	Employees become so secure in their work that they take shortcuts in performing their tasks, fail to perform to the same quality as before or become unaware of deficiencies	E1, E2	25
	Process	Lack of monitoring	Loss of control and visibility of the usage, enterprise data transmission on a personal device	E1, E3, E4, E5	50
		Unaware of policy	Employees being ignorant or uninformed about BYOD security policies	E1, E2, E5	37.5
		Lack of enforcement	Policies exist in the organization without being imposed on employees	E1, E2	25
	Technology	Lack of antivirus	Poor or non-existent virus protection	E1, E3, E4, E5, E6	63
		Lack of protection mechanism	Having less or no measures to secure the devices	E1, E2, E3, E5	50
		Mixed personal and organizational data	Both organizational and personal data is stored in personal devices	E1, E3, E4, E8	50
		Malware	Malicious application that is able to sniff, modify, or steal and compromise trusted applications on the device	E1, E6, E7	38
		Device loss	Physical loss or theft of a device	E2, E5	25
		Shadow IT	Information technology (IT) solutions and systems without prior explicit organizational approval	E3, E5	25

devices in hand, it is quite easy for someone sitting in a private meeting to secretly record and disseminate confidential information over the Internet instantaneously. *"These acts can cause data leakage and data manipulation" (E4). E5* highlighted that:

Some employees download minutes of meetings that are classified as 'Secret' into their laptops, tablets, or mobile devices and sometimes it becomes a risk, when someone intentionally screenshots important parts of the document and shares it in social media. There are instances where the data is wrongly shared to the unintended person or in the wrong chat group. With huge amount of data in devices, it is quite difficult to identify what the data is about as the files are all numbered, and we do not really rename them or separate them from other data in the devices. This can jeopardize the Confidentiality, Integrity and the Availability (CIA) of the data and this is a huge challenge." E3 further cautioned that *"the challenge is not only to ensure that government data is not leaked, but also our personal data, especially when we store our IDs, passwords and personal pictures in devices. In some cases, personal data when compromised can be sold online for a hefty price.*

b. **Complacent behavior**

Most employees never secure their devices, according to the participants. Employees, according to E1, are rather blasé regarding device security. "*Sometimes, they will follow the rules when it's the first time, later when they are busy or want to complete things fast, they will skip security procedures" (E1). E5* cautioned that *"threat does not necessarily come from technology alone, it can also come from humans, especially the insiders. Insiders are the ones who can steal the data" (E5).*

5.1.1.2 Process Risks

a. **Unaware of risk**

Experts cautioned that employees' seeming ignorance of security policies may heighten firms' security risk exposure. "They just assume they can bring their personal device and use it for official work purposes" (E5). "Some employees are unaware of the security classification of documents and how it should be handled or stored as stipulated in the Official Secrets Act 1972. Additionally, if there are any organizational data in the devices that should be deleted, employees must follow proper sanitization procedure, else the data could be retrieved back" (E1, E2). "Many do not adhere to the policy as some employees still send classified document via emails" (E3).

b. **Lack of enforcement**

The experts also discussed the flaws that exist in the policy enforcement in agencies. "The public sector uses the ICT security policy that has a general clause on BYOD security, which is quite inadequate to manage BYOD in the office" (E1). Because of this, organizations may find themselves unable to impose Bring Your Own

Device policies on their staff. "Even though, it is stipulated in the security policy that employees cannot use personal devices for official work purposes without permission, many still bring due to lack of enforcement and monitoring". E2 warns that it is almost commonplace for everyone to use their own devices for work purposes. E1 indicated "it is also difficult to enforce policies and rules or security controls on employees as the devices belongs to them. That becomes our weakness. It is different if we have a strong policy and good enforcement, then it shall be a pre-requisite to adhere to the policy if they choose to use their devices for work purposes". In addition, E1 underlined that the lack of a mandate is a significant problem within the organizations. "When there is lack of mandate and commitment from top management, enforcement becomes an option for the rest of them in the organization. It is difficult to push security controls from bottom to up."

The majority of experts concurred that some of these firms lack any monitoring or control system for BYOD activities and security issues; *"... we do not monitor closely employees' BYOD device activities. The organizations don't know what information actually resides in their devices, when that should be a concern. We do not have the tools and software to monitor them..." (E1, E3, E4, E5).* The device ownership conflicts with monitoring of devices... as reported by *E5, "if it's our office computer, we can still check and extract the data but with employees' devices, we are unable to do that".*

5.1.1.3 Technology Risks

a. **Shadow IT**

Shadow IT, or the usage of a plethora of mobile devices and mobile operating systems without the explicit agreement or knowledge of the IT department, has made it difficult for organizations to govern them. As E3 demonstrates, asset and network management become increasingly difficult, *"it is difficult to trace the devices or trouble shoot, as sometimes they may congest the network and they are not visible to the IT administrators". "The employees also perform activities without the knowledge of the IT department" (E5).*

b. **Lack of antivirus**

Additionally, organizations confront issues in safeguarding these devices such as "...for the office computers, we have centralized antivirus, valid software licenses and scheduled maintenance unlike in personal devices, which has almost none of these. Hence, the devices become a threat when connects to the secured office network. The malware can spread quickly via the network and quickly infect the network to target websites and infrastructures" (E1, E3, E4, E5, E6).

c. **Lack of protection mechanism**

Lack of vigilance makes devices and organizations susceptible to mobile threats. "The protection mechanism for the devices is not as tight as the computers provided by the organization" (E5). "Before, we know that our data is in our computer, in our office rooms, protected by secured doors of the organization, but now with BYOD, we are already out of the security parameter" (E2). "If one keeps classified documents in digital format in their devices, it should be handled like a physical classified document, where it has security procedures that one should abide. Similar to physical documents, they should have controls like using passwords and encrypting the hard disk. These should be considered as a pre-requisite if one wants to practice BYOD" (E1). "If the organization has some tools that can be used to scan and check the security features of the device, before releasing internet protocol (IP) to it, then employees will be encouraged or forced to install some security controls in their devices" (E3).

d. **Mixed personal and organizational data**

As mobile devices become the primary computing platform for many enterprise employees, a large amount of corporate data, including work files, emails, and organizational records, will reside on the devices.

We do not know about the data that is stored in the devices; it could be confidential information about the organization. That information is our asset, it could be customer information, business strategies or information on potential clients. Data is the new oil" (E8). Employees keep documents belonging to the organization in their devices and we do not have any tools to monitor those" (E4). In the device, it is actually hard to separate personal and organizational data. It is also a challenge to impose security controls on those information" (E1). Previously, with organization owned computers, all data belongs and will be returned to the organization when the employee leaves, but now with BYOD, it is different as device belongs to the employee. When employee leaves the organization, they bring all the data with them (E3).

Even after employees leave the firm, there is still a risk that they will bring their personal device to a new job and leak or mistakenly reveal critical information with their new employers. Furthermore, sending laptops, smartphones, and tablets to vendors for repair or replacement may undermine the security of sensitive data, especially if suitable protection methods are not implemented. *"Some may end up with data loss or manipulated data" (E5).*

e. **Device Loss**

Theft or loss of devices adds another layer to the risk complexity. *"This becomes a reason for data leakage, when the sensitive information inside gets into other people's hands" (E2).* The significance of the information stored on these devices, combined with the fact that they do not use encryption, makes them desirable targets.

f. **Malware**

All eight experts concurred that malware poses a substantial threat to the organizations. E1 pointed out, "When there is another network gateway using their mobile network that provides access to the entire Internet, the security threats will increase. Though the security team has lined up all the security applications and firewalls, it is pointless when employees use the mobile data and hooks up on to the internal network, that brings the uncontrollable cyber-attacks like ransomware". These devices become infected with malware as a result of installing a malicious program or defective device security, and when they connect to the organization's network, the entire network is vulnerable to a data breach. "As we have a cyber-forensic team, we know that there are lots of threats, lots of vulnerabilities in the phones itself" (E6). To get access to systems and data, hackers disguise or spoof the communications from unknown sources as coming from known, trustworthy sources. This activity also involved the spread of malware in order to launch a cyberattack against organizations. "They send email that make employees click on unknown links and then malware propagates into the network. Quite often, we get ransomware through email" (E6, E7).

5.1.2 BYOD Security Risks Assessment

Security risk assessment is a crucial component of risk management, which includes both qualitative and quantitative ways to address information security vulnerabilities [122]. The ability to identify risk that threaten the organizations' IT systems and data is critical to establishing mitigation strategies [119]. The findings should lead and identify the relevant management action and priorities for managing information security risks as well as implementing appropriate controls to mitigate these risks. Existence of security risk has strong influence on the understanding or the acceptance of security beliefs which will ultimately influence security policy compliance [9]. Understanding the risks and managing them appropriately will enhance the organizations' ability to achieve their information security goals [70]. By considering risk and identifying factors of security compliance based on risk, the likelihood of achieving security goals are improved, consistent and enables better compliance management [70].

5.1.2.1 Findings of BYOD Security Risks Assessment

Experts were asked to conduct risk assessment in this study based on a list of identified security risks (Table 5.2) determined by performing a systematic literature review on 17 existing literatures that were relevant to employees' BYOD security policy noncompliance behavior. Experts were asked to evaluate the impact of the risks based on "Hardware", 'Software', 'Information' and 'Services'. Each expert articulated that only 'Information' and 'Services' are valuable assets that matters for organizations in the case of personal devices. This is because "Hardware" and "Software" are elements that belong to employees and have less of an effect on organizations hence those categories were omitted. As a result, the impact was assessed based on the degree of overall damage or loss that could arise from the exploitation of security vulnerabilities in the "Information" and "Services" categories of the organization.

Table 5.2 Security risks from employees' BYOD security policy non-compliance behavior [85]

People		Process		Technology	
		Policy		*Device management*	
CB	Complacent behavior	LP	Lack of policy/ guidelines	SM	Stolen/misplaced
MP	Misuse of privilege	UE	Unfair enforcement	UM	Unreported missing devices
VP	Victim of phishing	LM	Lack of mandate	UU	Unauthorized usage
SI	Sharing information online	UP	Unaware of policy	PO	Personal and organizational data stored together
FC	Failure to configure	VI	Various interpretation due to ambiguity	SS	Small screens leading to phishing
ST	Find security procedures troublesome/intrusive	VS	Very strict policy	SP	Sharing passwords
CO	Careless due to over confidence	*Education*		PM	Lack of protection mechanism (screen lock/password)
		LT	Lack of training	MA	Malware attack
		LK	Lack of knowledge/ skills/self-efficacy	LA	Lack antivirus
		UR	Unaware of risk/threat severity	*Technical issues*	
				AT	Unable to track all devices (shadow IT)
				VO	Various operating systems
				RW	Connecting to risky, unknown Wi-Fi

The risk assessment forms and ratings were adapted from the Information Security Management System (ISMS) Risk Assessment Methodology [65] (Please refer Appendix B). The risks were classified "Low", "Medium" or "High". The more severe the repercussions of a threat, the higher the risk. The implications might be referenced to the Risk Rating Table (please refer Appendix C). The likelihood assessment analyses the possibility of a threat occurring. The greater the possibility of a threat occurring, the higher is the risk. It can be difficult to reasonably quantify likelihood for many parameters; hence, to simplify relative likelihood would be good using High/Medium/Low as measurement. In order to establish an overall estimated degree of risk, it is necessary to determine the impact and likelihood of each risk that has been identified. Both Figs. 5.1 and 5.2 show the number of experts that evaluate each risk whether it is "Low', 'Medium' or 'High" for both "Information' and 'Services' categories.

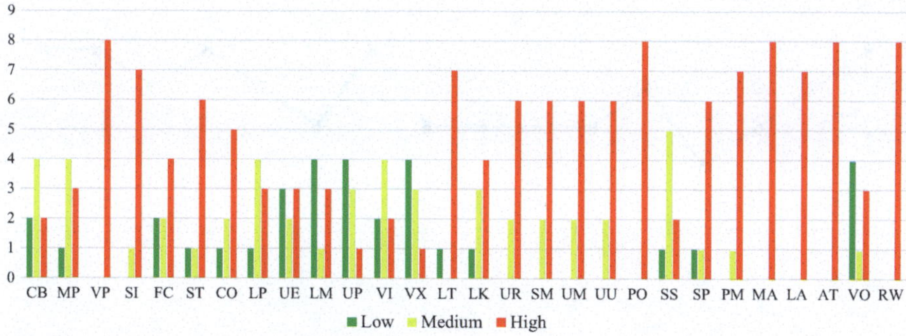

Fig. 5.1 Risk assessment for the category "information" [84]

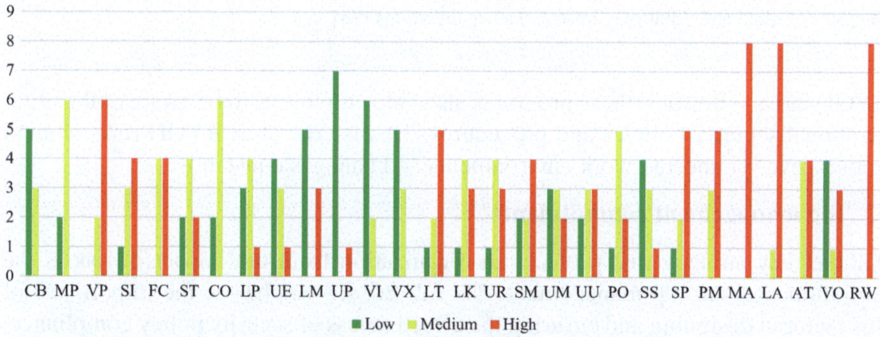

Fig. 5.2 Risk assessment for the category "services" [84]

In this risk assessment procedure, each risk rated "High" by five or more (majority) of the total eight experts is deemed critical. Figure 5.3 depicts the identification of sixteen risks that were evaluated "High" in both categories.

5.1.2.2 Control Measures that Mitigate Risks

Experts advocate these controls to prevent BYOD cyber risks, and by integrating them into organizational security processes, BYOD security policy compliance can be strengthened. Figure 5.3 illustrates the control measures identified for each risk.

a. **Security Training and Awareness (SETA)**

All experts suggest SETA as an important strategy to be employed to reduce the identified risks levels in the organization. They agree that SETA programs are integral to building a culture of security in the organization. *E6, E8* recommended cybersecurity competency development and assessment to the workforce in order to address

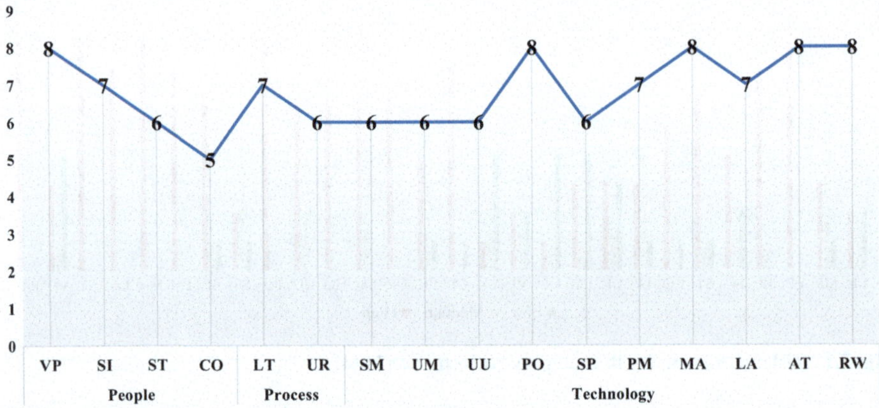

Fig. 5.3 Critical risks that may impact the organizations [84]

BYOD threats. Besides, these programs also help employees to be aware of the organizational security policies and procedures. This is critical as BYOD risks abound in these ever-connected work environments and changes constantly.

b. **Top management commitment**

Another key strategy that protects an organization from the impact of risk is the commitment from top management. *E1, E2, E6, E8* emphasize the importance of this factor in designing and ensuring the effectiveness of security policy compliance. Even though BYOD is practiced with a bottom-up approach, successful information security implementation requires top management to demonstrate effective leadership by setting a *"tone at the top"*. *E4* stressed that management security meetings to report and evaluate information security performances should be held on a continuous basis. Although this strategy could not be identified as a mitigation strategy for most of the risks, experts claim that without the management factor, all information security efforts can be in vain. *E1* and *E2* also exemplified that *"if there is a pyramid of factors, this factor should be positioned at the top to indicate its importance"*.

c. **Policy**

Information security policies play an important role in an organization's overall security posture. Defining and specifying good security behavior in security policies holds individuals accountable for their non-compliance actions. Security experts believe that having a good security policy can help to chart a roadmap to employees of what to do and when to do it. This can also help to coordinate and communicate security measures across the organization.

d. **Technical countermeasure**

All experts suggest technical solutions to minimize most of the BYOD risks impact in organizations. Device monitoring and management, secure authentication and encryption are among the technical solutions that can be implemented parallel to the organizational security measures to ensure comprehensive cyber protection strategies.

5.2 Discussion on Qualitative Findings

The study finds that the BYOD features can be important innovation drivers and at the same time pose significant risks to the organizations [13, 16, 85, 114]. The ubiquitous nature of the devices and relative insecurity of mobile devices makes it a target for cybercriminals, malware and other security threats [69, 87]. Their portability nature also means an increased risk of devices becoming lost or stolen and putting a huge risk on government data. When used in public areas and connecting to unsecure public Wi-Fi networks, employees are at a greater risk of being hacked.

According to security experts, technology-related risks are the most critical, followed by people-related risks and then process-related risks. This may be attributable to their access to statistics on the frequency of attacks originating from firewall services or virus infections, which heightens their awareness of the risks posed by technology.

Mobile devices present unique risks that can lead to adverse outcomes, and explains the need for users to take special measures to reduce or prevent them [115]. The same security measures that organizations use for workstations and servers may be inadequate for mobile devices as they are not built with robust security features unlike desktops. Because of this fallacy, mobile devices may not be protected by firewalls, encryption, or antivirus software.

The findings confirmed that employees were disclosing information about work and classified information via mobile devices might jeopardize organizational information security. Furthermore, the growth of social media applications and it being readily available in personal devices causes employees to disclose information online easily, either inadvertently or intentionally [1]. This allows phishers to obtain information through publicly available organizational and employees data before launching attacks [87]. Additionally, intentionally or unintentionally sharing information online by negligent employees leads to data leakage which will result in costly data breach and reputational damages [53].

The findings also reveal that data ownership is a challenge as organization need to define who owns which data, while finding a balance between data protection and employee privacy. Organizations should prepare an exit strategy to collect organization-owned data and information from the departing employee to prevent data theft. This move is crucial as employees are constantly assessing, transferring data and intellectual property without any auditing or tracking by IT department and often against corporate governance.

Devices that are not secured with any protection mechanism pose severe risks to the organization. Employees must be trained to ensure that they have adequate protection measures in their personal devices like installing antivirus, enabling password or phone locking mechanisms and regularly updating the devices. This is highly recommended since regular updates help counter security vulnerabilities and other issues.

It is also very evident that dealing with difficult employees is a big challenge that organizations have to face even with the most sophisticated technology [50]. Employees' lack of cybersecurity awareness creates a fertile ground for hacks [12] as it takes only one careless employee to leave the whole organization's network at risk. However, today, the parameters have shifted and there are a vast number of dissimilar endpoints, with more employees using smartphones for work purposes.

This study also found most of the organizations are unable to monitor their employees' BYOD security behavior. On the other hand, if they do monitor and manage every aspect of incoming devices, this could be intruding employees' privacy. BYOD monitoring can be a double-edged sword for organizations if employers don't incorporate a mechanism that distinguishes corporate content from personal information privacy and put employees at risk.

Handling employee security behavior issues might be more intricate as these devices belong to the employees and traditional 'rules' or mandates for enforcing security simply may not work and may result in the legal consequences if device monitoring is not executed properly [105]. The shift of device ownership has many complex and hidden implications to be considered before implementing any security strategy. Organizations need to address policy issues by creating a balance of freedom, control, compatibilities and security [107] while the organizational security objective remain paramount. If a policy includes rigorous controls, employees will either discover a means to evade it or it will impede the usage of BYOD by employees, resulting in a loss of business value.

The critical risks deserve attention to prevent compromised information security efforts. Experts suggest four main control measures to mitigate the identified risks. Notably, these measures are not necessarily quick fix to closing security gaps, rather, finding the right balance between level of security and acceptable risks. This can also be a way to motivate employees to protect the security of organizational information from potential risk [76, 98]. These risk reduction approaches are vital in ensuring the confidentiality, integrity and availability of information stored in devices [98]. The most suggested strategy is policy, followed by SETA, technical countermeasures and lastly management commitment. This could be due to the fact that security managers rely on approaches that are easier to implement and within their power. Roping in top management commitment is a challenge in certain organizations, especially those for which IT is not their core business, information security decisions and responsibilities are assigned to lower-level IT managers, who are assumed to be the best candidates to ensure top notch security. Lack of support from top management can falter organizational missions and risk the safety of the public data. Consequently, the managers probably viewed the rest of the control measures especially the technical countermeasure as an important means of controlling and monitoring employee BYOD behavior. However, this can be futile if for example, there are many missing devices that are not formally reported to the management, making remote wiping tools or any other device management tools to be useless. Hiding an incident such as this may increase the damage caused.

A comprehensive BYOD security policy serves as a basis for solid enforcement and a backstop against legal issues for organizations. A policy that is straightforward and effectively conveyed may assist organizations in managing the use of personal devices [25]. SETA empowers employees with skills to safeguard their devices from cyber threats [13, 31]. Success or failure of an organization security efforts relies on these insiders diligence and knowledge [91, 94]. SETA must consider two groups of employees: those who are adept users of their device and readily incorporate its use into their work and those who face a much steeper learning curve. Organization should also be vigilant on those who are comfortable with the technology but unfamiliar with working within the constraints of a corporate policy governing its use.

Focusing on technology factors and overlooking the human factor may lead to dramatic consequences for the organization. The human aspects of information security must be given priority to transform employees from being the weakest link into the line of defense. This can be achieved by having effective information security culture that encompasses both technical and human issues [6, 44]. The human factors are not solely about the employees' security education but also on social and environmental forces that may influence employees' security behavior in organizations as well. Influence of people such as peers and family, managers who constantly form a social influence within the workplace may shape their security behavior. Management plays an important role in setting the information security goals and objectives of the organizations.

The control measures generated from this study can be used as factors to improve BYOD security policy behavior because their implementation helps to reduce or eliminate security threats (Fig. 5.4). The security of personal devices must begin with the visibility of threats. One cannot implement strategies without understanding the nature and severity of the risk they face.

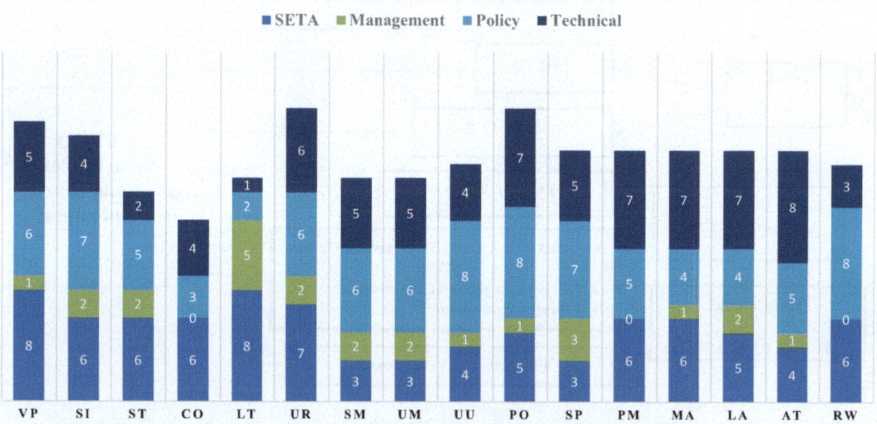

Fig. 5.4 Control measures to mitigate risk recommended for each critical risk [84]

The 'Policy' strategy can be interpreted through the Organizational Control Theory [21, 71]. The analysis of the components of control by specifying and evaluating security rules, has the capacity to provide insight on the impact of management policies and procedures on individual compliance [21]. A basic and a well-articulated policy may help controlling the security risk from personal devices in organizations [25].

Control measures such as top management commitment and technical countermeasure (interpreted as device monitoring) can be explained as security culture [34, 52]. Security culture reflects the values and beliefs of information security shared by all members at all levels of the organization.

SETA is seen as an independent construct that can be connected to self-efficacy. SETA is described as a platform for organizations to instill and cultivate information security awareness, educate users on the necessity of information security, and train personnel to assume information security responsibilities.

The mapping of each control measure to the conceptual research framework is shown in Fig. 5.5. The qualitative research findings are important as they provide the empirical justification to the conceptual research framework and will be validated during the quantitative phase. Additional factors in the conceptual framework (Psychological ownership and BYOD IT support) are maintained to be tested in the next study as they are less-researched and have the potential to influence BYOD security policy compliance.

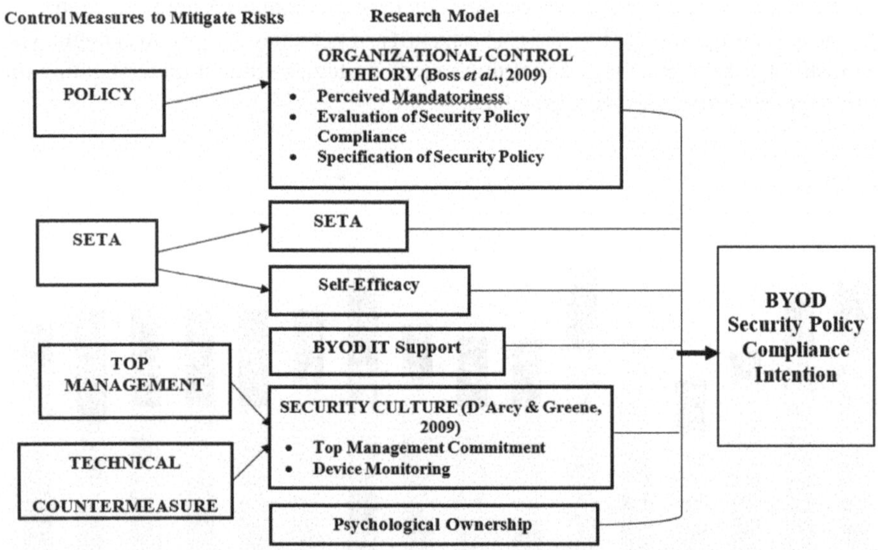

Fig. 5.5 Mapping of control measures to the BYOD security policy compliance framework

5.3 Quantitative Research Findings

Following the qualitative research, a survey was distributed online to CNII agencies' employees to obtain their perception on BYOD security policy compliance factors. The process begins with data collection and followed by data preparations, including data editing, coding and screening, before the PLS-SEM assessment is performed. General description of the survey respondents is presented followed by the results of Confirmatory Factor Analysis (CFA) which covers the assessment of reliability and construct validity. Structural model results of the hypotheses put forth in Chapter Three and the predictive relevance of the research model are presented in section four.

5.3.1 Preliminary Analysis

Preliminary data analysis is important to ensure that the quantitative data used in the research are in order before proceeding with PLS-SEM analysis. Data preparation involves editing, coding and data entry to ensure the precision and adequacy of the data that will be analyzed.

5.3.2 Data Structure

The process begins by indicating the name of the variables involved. This research survey uses close ended questions which comprises thirty-seven items for the independent variables (IV) and three items for dependent variables (DV) and ten items for demographic questions.

The next step is the coding process, where categories are assigned with numbers or symbols to facilitate the measurement of items into a certain number of groups or categories. For independent variables, the responses are pre-coded as 1–7 and 1–5, with 1 indicating strongly disagree, 5 and 7 indicating strongly agree.

5.3.3 Data Cleaning

Data cleaning involves the process of checking for blank responses and missing values.

5.3.3.1 Blank Responses

The first step in data cleaning is checking for the blank responses among the collected questionnaire and Microsoft Excel was used to conduct the procedure. Some of the respondents attempted the survey, but they left it halfway after completing the demographic section. Cases with blank responses were deleted from the sample.

5.3.3.2 Missing Values

According to Hair et al. [55], missing data occurs when respondent does not answer one or more questions, either intentionally or unintentionally. The effect of missing data is that it can diminish the statistical power of a study and may generate biased estimates that results. As such, the online survey form was set to prevent respondents from skipping questions to ensure all questions are answered before submitting them.

5.3.4 Data Normality

Mardia's multivariate kurtosis is used to check the normality of the data. Cain et al. [23] and Hair et al. [55] suggested that the multivariate skewness and kurtosis are measured using an online statistical analysis. The results indicate that the data is not normally distributed, as shown by the skewness ($\beta = 41.723$, $p < 0.01$) and kurtosis ($\beta = 203.747$, $p < 0.01$). The tests show that the normal distribution assumption of the data is violated and there is a need to use Smart PLS, which is a non-parametric analysis tool.

5.3.5 Profile of Respondents

Once the data is prepared, the next stage is to run descriptive analysis. Descriptive analysis was performed, before administering structural equation modelling partial least square (PLS-SEM) to test the research model. The demographic details of the respondents are highlighted in Table 5.3. In terms of gender, there are 138 male respondents (39.9%) and 208 females (60.1%). In terms of age, 25 respondents (7.2%) are between the age of 21 and 30 years old group, 162 respondents (46.8%) are in the 31–40 years old group and 133 respondents (38.4%) are in the 41–50 years old group. 26 respondents (7.5%) are in the 51–60 years old group. Majority of the respondents (32.9%) have 11–15 years of working experience, followed by 27.5% have 16–20 years of working experience. The smallest group (2.6%) has less than a year work experience. In terms of education level, the distribution of the data is shown in Table 5.3.

Table 5.3 Participants' demographic information

Measure	Items	Frequency	Percent (%)
Gender	Male	138	39.9
	Female	208	60.1
Age	21–30 years old	25	7.2
	31–40 years old	162	46.8
	41–50 years old	133	38.4
	51–60 years old	26	7.5
Work experience	Less than a year	9	2.6
	Between 1 and 5 years	28	8.1
	Between 6 and 10 years	44	12.7
	Between 11 and 15 years	114	32.9
	Between 16 and 20 years	95	27.5
	Between 21 and 25 years	23	6.6
	Between 26 and 30 years	21	6.1
	More than 30 years	12	3.5
Education	High school	36	10.4
	Diploma	73	21.1
	Advance diploma	7	2.0
	Degree	138	39.9
	Master	88	25.4
	Doctorate	4	1.2
Total		346	100

5.3.6 Descriptive Statistics on BYOD Usage

Besides basic demographic information, a few questions were also asked to understand the type of devices employees use for work related tasks and the type of work they perform with their devices. Majority of the employees (300 respondents = 86.71%) use their smartphones and a total of 242 respondents (69.94%) use their personal laptop for work related tasks. Only 41 employees (11.85%) use tablets for BYOD purposes.

Almost all the employees examined, 98.13% read emails on their mobile devices, followed by 91.91% who read work-related documents such as reports, slides, and working papers. The majority of employees, 80.35%, use their personal devices to compose professional emails, while 62.14% also use their devices to compose reports and documents.

5.3.7 Common Method Bias (CMB)

Common method bias issues arise especially data is collected from a single data source. Therefore, before proceeding with CFA, the common method bias was evaluated using Harman's single factor test. CMB is described as 'variance that is attributable to the measurement method rather than to the constructs the measurement represent' [88], which could be problematic. It is a systematic error variance that is shared among variables which are measured from the same source or methods [95]. Testing on the threats of CMB is crucial when the data is collected via self-administered questionnaire and particularly when the predictor and criterion variables are obtained from the same person [89]. CMB may jeopardize the validity of the constructs and introduces systematic bias into an investigation [111].

In order to reduce the CMB, this study opts for both procedural and statistical remedies before and after data collection. For statistical remedies, Harman's single factor test and full collinearity were utilized to examine the CMB. Harman's single factor is based on the assumption that if a significant amount of CMB is present, a factor analysis of all the data will result in a single factor that accounts for the majority of the covariance in the variables. An unrotated single factor analysis of all study items explained less than 50% of the variance. Given that the value is 35% of the total variance, CMB was not viewed as an issue.

5.3.8 Confirmatory Factor Analysis—Measurement Model

The research framework for this study is tested using Smart PLS 3.0 [96]. Based on the two stage analytical procedures suggested by [14], this study examined the measurement model (validity and reliability of the measures) and the structural model (testing the hypothesized relationships).

The research framework was developed based on the literature review with a theoretical background, which is why confirmatory factor analysis (CFA) was considered to be a suitable approach for this research. CFA is a multivariate statistical analysis technique that maps the measured variables onto the factors developed through the theoretical understanding of the respective domain. CFA investigates whether the measured variables are consistent with our understanding of the variables and factors in the research model. The following subsections discuss the guidelines used to access the measurement model.

5.3.8.1 Internal Consistency Reliability

The first assessment conducted is to evaluate the internal consistency and reliability. Two tests were performed using Cronbach's Alpha and composite reliability index. It is imperative to ensure the reliability and validity of the questionnaire used in survey methods. The concept of validity implies that care is taken to ensure that the questionnaire assess what it claims to assess.

According to Table 5.4, the Cronbach's Alpha values in this study vary from 0.837 to 0.949, which meets the 0.7 threshold proposed by Hair et al. [55]. There have been arguments on the use of Cronbach's Alpha (α) as a tool to measure reliability [54]. Therefore, due to its deficiency, the composite reliability index is used instead as an alternative reliability test [79]. Composite reliability is a more rigorous assessment of reliability than Cronbach's Alpha because it can determine whether the specific indicators adequately represent the constructs [26, 46]. As such when assessing the measurement model, construct reliability was evaluated based on composite reliability (CR) and convergent validity. According to guidelines, composite reliability with a measure more than 0.7 indicates acceptable internal consistency for the model. Table 5.4 demonstrates that all constructs have composite reliability values higher than the threshold of 0.7 [55].

5.3.8.2 Indicator Reliability (Outer Loadings)

Once the reliability of each internal consistency reliability has been confirmed, the indicator reliability is measured. As shown in Table 5.4, all items have satisfactory indicator reliability ranging from 0.670 to 0.949. Since the values meet the threshold values, therefore none of the items were dropped.

5.3.8.3 Convergent Validity

Convergent validity refers to the extent to which individual indicators reflect the constructs in comparison to indicators measuring other constructs [116]. Convergent validity is achieved when items loading and average variance extracted (AVE) is measured. The AVE value should be greater than 0.70 and 0.50 [55]. The AVEs for all constructs in the model were above the cut off value of 0.50, indicating adequate convergent validity (Table 5.4). The lowest AVE value reported is for Psychological Ownership (0.608), followed by Specification of Policy (0.704) and Device Monitoring (0.754). Rest of the constructs, Evaluation of Security Policy, Perceived Mandatoriness, BYOD IT Support, Security Education Training and Awareness, Self-Efficacy and Top Management scored a higher value of AVE, which explains more than 80% of the total variance. The results show that the measurement model demonstrated adequate convergent validity.

5.3.8.4 Discriminant Validity

Discriminant validity is the degree to which items differentiate among constructs or measure distinct concepts and is measured by determining the correlations between the measures that have the potential to overlap [93]. Henseler et al. [60] proposed

Table 5.4 Assessment of construct reliability and convergent validity

Latent variable	Item	Outer loading	Cronbach alpha	CR	AVE
Specification of security policy (SOSP)	SOSP1	0.842	0.860	0.905	0.704
	SOSP2	0.875			
	SOSP3	0.809			
	SOSP4	0.828			
Evaluation of security policy (EVSP)	EVSP1	0.891	0.949	0.963	0.867
	EVSP2	0.949			
	EVSP3	0.949			
	EVSP4	0.936			
Perceived mandatoriness (PM)	PM1	0.841	0.900	0.930	0.769
	PM2	0.879			
	PM3	0.894			
	PM4	0.893			
BYOD IT support (ITS)	ITS1	0.883	0.927	0.948	0.821
	ITS2	0.935			
	ITS3	0.893			
	ITS4	0.912			
Security education training and awareness (SETA)	SETA1	0.857	0.941	0.955	0.810
	SETA2	0.922			
	SETA3	0.923			
	SETA4	0.907			
	SETA5	0.889			
Self-efficacy (SE)	SE1	0.893	0.897	0.936	0.830
	SE2	0.912			
	SE3	0.926			
Top management (TM)	TM1	0.909	0.891	0.932	0.821
	TM2	0.909			
	TM3	0.899			
Device monitoring (MON)	MON1	0.829	0.837	0.902	0.754
	MON2	0.909			
	MON3	0.865			
Psychological ownership (PSYO)	PSYO1	0.700	0.892	0.915	0.608
	PSYO2	0.670			
	PSYO3	0.796			
	PSYO4	0.760			
	PSYO5	0.839			
	PSYO6	0.847			
	PSYO7	0.828			

(continued)

Table 5.4 (continued)

Latent variable	Item	Outer loading	Cronbach alpha	CR	AVE
Intention to comply (INT)	INT1	0.946	0.942	0.963	0.896
	INT2	0.956			
	INT3	0.938			

Table 5.5 Assessment of discriminant validity using HTMT criterion

		1	2	3	4	5	6	7	8	9	10
1	EVSP										
2	INT	0.261									
3	ITS	0.460	0.256								
4	MON	0.634	0.367	0.666							
5	PM	0.387	0.435	0.38	0.41						
6	PSYO	0.113	0.454	0.161	0.214	0.318					
7	SE	0.465	0.478	0.7	0.597	0.506	0.34				
8	SETA	0.604	0.338	0.725	0.76	0.383	0.21	0.57			
9	SOSP	0.657	0.347	0.569	0.514	0.723	0.252	0.655	0.627		
10	TM	0.536	0.38	0.573	0.788	0.417	0.191	0.532	0.74	0.541	

Note HTMT < 0.85

the Heterotrait-Monotrait Ratio (HTMT) as a criterion for assessing discriminant validity at the final stage of assessment and updated by [47]. HTMT is used over the traditional [46] as it offers the best balance between high detection and low arbitrary violation rates. The HTMT values should be ≤ 0.85 the stricter criterion and the mode lenient criterion is it should be ≤ 0.90. If the HTMT values are greater than 0.85, it shows the evidence of discriminant validity issues. As indicated in Table 5.5, the HTMT values were all less than the stricter criterion of 0.85; therefore, it can be concluded that respondents recognized that each construct is unique. Together, these validity tests have demonstrated that the measurement items are both valid and reliable.

5.3.9 Higher-Order Construct (HOC)

In the current study, the construct of security culture is conceptualized formatively. As discussed in previous section, in this study, security culture is measured using two dimensions' constructs, namely top management and device monitoring. The assessment of HOC is tested using variance inflation factor (VIF) values which is ranging from 1.590 to 2.883, less than threshold score of 3.3 indicate that multicollinearity is not a concern in this path model [42].

Table 5.6 Summary of measurement model assessment

Validity type	Criterion	Results
Internal consistency reliability	Cronbach's alpha (CA)	The CA value is acceptable if greater than 0.7 [55]. In this study, CA values ranged from 0.837 to 0.949
	Composite reliability (CR)	The CR value for constructs ranged from 0.902 to 0.963, which exceeded the recommended value of 0.7 [55]
Indicator reliability	Indicator loadings	Values should be higher than 0.700. All indicators are above 0.700
Convergent validity	Average variance extracted (AVE)	The AVE value for each construct ranged between 0.608 and 0.896. Values obtained are higher than threshold value 0.500
Discriminant validity	HTMT	HTMT values are lower than 0.85

In addition, the significance and outer weights of the constructs further validate the data, as all of the lower-order constructs (LOC) are significant at 0.000 [57] as shown in Table 5.7. It can, therefore, be concluded that both reliability and validity requirements are met for this study. Next, the data can be further analyzed for structural measurement.

Table 5.7 Assessment of higher-order construct (HOC)

HOC	Dimensions for LOC	Outer weights	Outer VIF	t-value	p-value
Top management (TM)	TM1 <- TOPM	0.389	2.589	67.406	0.000
	TM2 <- TOPM	0.348	2.818	58.485	0.000
	TM3 <- TOPM	0.367	2.498	57.496	0.000
Device monitoring (MON)	MON1 <- MON	0.413	1.590	36.024	0.000
	MON2 <- MON	0.389	2.883	69.222	0.000
	MON3 <- MON	0.352	2.500	36.094	0.000

5.3.10 *Structural Model*

Composite reliabilities, discriminant and convergent validities as stated in Table 5.6, have fulfilled the adequacy of psychometric properties within the designed structural model. The structural model was assessed after the establishment of the measurement model to determine the model's capabilities to predict one or more target constructs [55]. Bootstrapping with 5000 samples was used to calculate the path estimate and robust t-statistics for the hypothesized relationships. PLS-SEM does not assume that the data are normally distributed, hence parametric significance tests cannot be used to determine the significance of coefficients such as outer weights, outer loadings, and path coefficients. PLS-SEM instead uses a nonparametric bootstrap approach to assess the significance of estimated path coefficients. The findings are presented in the following section.

5.3.10.1 Assessment of the Structural Model for Collinearity Issues

The first step in the structural model is to examine collinearity issues. It is to prevent collinearity issues between the constructs before performing a latent variable analysis in the structural model. The collinearity is measured by measuring the VIF value. The threshold value for the assessment is 5 [56] or 3.3 [42]. In this study, as presented in Table 5.8, all the inner VIF values for the constructs vary from 1 to a maximum 1.848, which is below the most adopted cut off value of 3.33 [42] denoting that multicollinearity is not an issue.

5.3.10.2 Assessment of the Structural Model Relationships Significance

To test the hypotheses', bootstrapping procedure is employed to produce results for each path relationship in the framework. Following the non-parametric guidelines, the hypotheses were assessed by bootstrapping with 5000 sub-samples (see Table 5.6). Bootstrapping in PLS is a non-parametric test which comprises repeated random sampling with replacement from the original sample to produce a bootstrap sample and to attain standard errors for hypothesis testing [56].

For this study, nine hypotheses are developed for the constructs. To test the significance level, t-statistics for all paths are generated using SmartPLS 3.0 bootstrapping function. The bootstrapping is set to 0.05 significance level, one tailed test and 5000 subsamples. The critical values for a significance level of 1% ($\alpha = 0.01$), 5% ($\alpha = 0.05$) and 10% ($\alpha = 0.1$) are 2.33, 1.645 and 1.28, respectively for the one-tailed test [93]. Based on the findings shown in Table 5.8, the value of path coefficients has a standardized value approximately between − 1 and + 1 (values from − 0.045 to 0.740). Estimated path coefficients near to + 1 indicate strong positive associations, whilst values closer to 0 indicate weaker relationships.

Table 5.8 Assessment of structural model with bootstrapping procedure

	Path hypotheses	Std beta	Std. error	t-value	p-value	Confidence interval	VIF	f^2
H1	SOSP → PM	0.679	0.082	8.157***	0.000	(0.542, 0.808)	1.547	0.497
H2	EVSP → PM	-0.045	0.087	0.425	0.335	(− 0.187, 0.010)	1.547	0.002
H3	PM → INT	0.179	0.08	2.175*	0.015	(0.049, 0.311)	1.353	0.033
H4	ITS → SE	0.537	0.067	7.965***	0.000	(0.422, 0.642)	1.848	0.274
H5	SETA → SE	0.165	0.072	2.278*	0.011	(0.046, 0.279)	1.848	0.025
H6	SETA → SECURITYCULTURE	0.740	0.04	18.361***	0.000	(0.669, 0.800)	1.000	1.198
H7	SE → INT	0.192	0.076	2.540***	0.006	(0.069, 0.319)	1.626	0.035
H8	SECURITYCULTURE → INT	0.133	0.086	1.621	0.053	(− 0.013, 0.270)	1.481	0.020
H9	PSYO → INT	0.303	0.066	4.430***	0.000	(0.192, 0.413)	1.139	0.114

Note * Significant at $p < 0.050$ (one-tailed test); *** Significant at $p < 0.001$(one-tailed test)

Table 5.9 Hypotheses testing result

Hypotheses		Decision
H1	Specification will be positively associated with the employees' perceived mandatoriness of the established set of security policies	Supported
H2	Evaluation of security policy compliance will be positively associated with the employees' perceived mandatoriness of the established ICT security policy	Not supported
H3	Individual's perceived mandatoriness of compliance with existing ICT Security Policy is positively associated with BYOD security policy compliance	Supported
H4	BYOD IT support is positively associated with self-efficacy	Supported
H5	SETA is positively associated with self-efficacy	Supported
H6	SETA is positively associated with security culture	Supported
H7	Self-efficacy is positively associated with BYOD security policy compliance	Supported
H8	Security culture is positively associated with BYOD security policy compliance	Not supported
H9	Psychological ownership is positively associated with BYOD security policy compliance	Supported

The bootstrapping results show significant paths pertaining to SOSP to PM ($\beta = 0.679$, $p < 0.001$) and PM is positively associated with INT ($\beta = 0.170$, $p < 0.05$), thus supporting hypotheses H1 and H3. However, results did not show such an effect for EVSP ($\beta = -0.045$, $p = 0.335$).

The results also show that the hypothesized relationships of ITS ($\beta = 0.537$, $p < 0.001$) and SETA ($\beta = 0.165$, $p < 0.05$) towards SE are positive and significant, providing empirical support for hypotheses H4 and H6. SE is also found to positively influence INT ($\beta = 0.192$, $p < 0.001$). Similarly, the hypothesized relationship between SETA and SECURITYCULTURE ($\beta = 0.74$, $p < 0.001$), thus supporting hypotheses H6. Surprisingly, SECURITYCULTURE is found to be non-significant to INT ($\beta = 0.133$, $p = 0.053$). Finally, the analysis confirms that PSYO has a positive effect on INT ($\beta = 0.303$, $p < 0.001$), indicating that psychological ownership will result in intention to comply. Summaries of the findings are shown in Table 5.9 and depicted in Fig. 5.6.

5.3.11 *The Coefficient of Determination (R^2)*

The next stage is to evaluate the framework's predictive accuracy through the coefficient of determination score (R^2). The R^2 computes the framework's predictive power, and the value ranges from 0 to 1, with a higher value indicating a higher level

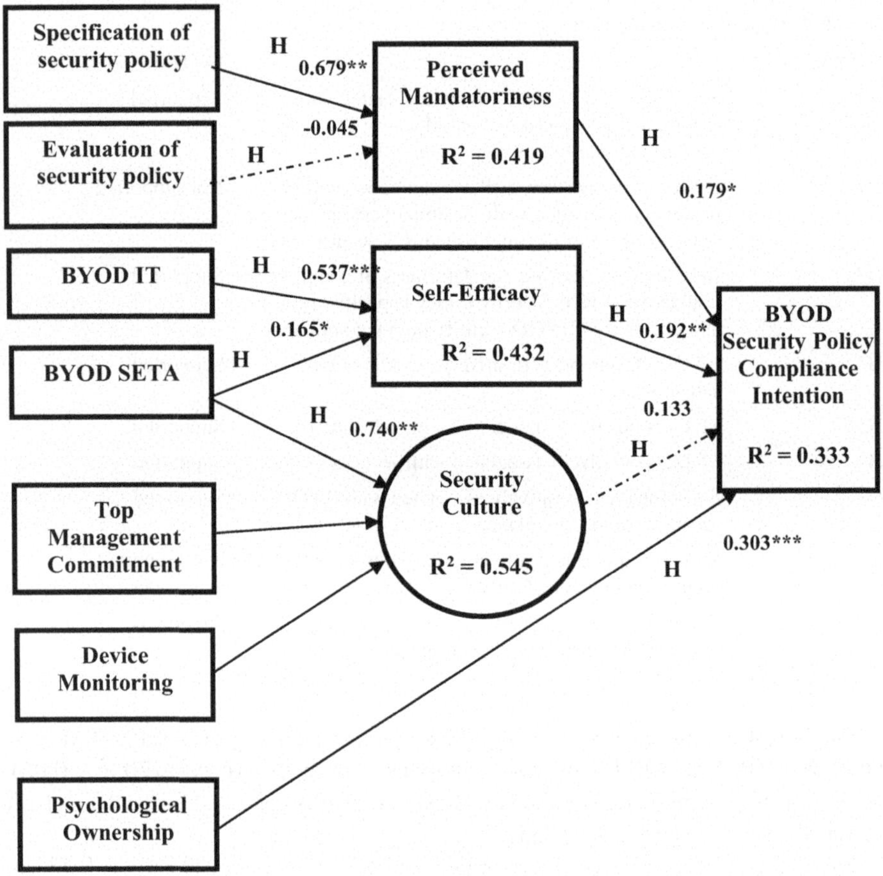

Fig. 5.6 Structural analysis

of predictive accuracy [55]. As there are various set of rules on the acceptable R^2 (refer Table 4.21, Chap. 4), this study follows the guideline by Cohen [30]. Values of 0.2 and above is considered high for studies dealing with human behavior. Overall, referring to Table 5.10, Specification of Security Policy and Evaluation of Security Policy explain 41.9% of the variance in Perceived Mandatoriness, which indicates a substantial level of predictive accuracy.

Meanwhile, BYOD IT Support and BYOD SETA explain 43.2% of the variance in Self-Efficacy, indicating a substantial level of predictive accuracy. Additionally, the highest variance (54.5%) is produced in Security Culture contributed by BYOD SETA, Top Management Commitment and Device Monitoring. Finally, Perceived Mandatoriness, Self-Efficacy, Security Culture and Psychological Ownership explain 33.3% of the variance in BYOD Security Policy Compliance Intention which also indicates a moderate level of predictive accuracy. It can be concluded that overall, the

Table 5.10 Predictive relevance

	Coefficient of determination, R^2	Predictive relevance, Q^2
INT	0.333	0.284
PM	0.419	0.312
SE	0.432	0.349
SECURITYCULTURE	0.545	0.456

framework is able to predict up to 33.3% of the factors influencing BYOD Security Policy Compliance Intention, which is satisfactory for a social science study.

5.3.12 *Assessment of the Effect Size* (f^2)

In this stage, the effect sizes (f^2) are evaluated. The f^2 computes the relative impact of a predictor construct on endogenous constructs. Besides reporting the p-value, both the substantive significance (effect size) and statistical significance (p-value) are crucial to be reported [108]. To measure the effect size, a guideline by Cohen [30] is followed. Based on the guideline, the values 0.02, 0.15 and 0.35 represent small, medium and large effects, respectively.

Looking at the effect size (f^2) values in Table 5.8, it can be observed that all the relationship possesses substantive impact on the outcome at various degrees. The f^2 of security culture (0.02), perceived mandatoriness (0.033), self-efficacy (0.035) and psychological ownership (0.114) indicate small yet significant effect size in generating R^2 for INT [28]. While, specification of security policy (0.497) has a substantial effect, evaluation of security policy (0.002) has a small effect on perceived mandatoriness (PM). BYOD IT support (0.274) has a medium effect, whereas SETA (0.025) has a small effect in explaining self-efficacy. SETA (1.198) has the biggest effect in explaining security culture.

5.3.13 *Predictive Relevance* (Q^2)

Lastly, a PLS predict technique was performed to determine the predictive relevance (Q^2predict) of the model. As recommended by Hair et al. [55], the predictive relevance of the research model was evaluated utilizing the blindfold technique. Based on blindfolding procedure, Q^2 evaluates the predictive validity of a large complex model using PLS. While estimating parameters for a model under blindfolding procedure, this technique omits data for a given block of indicators and then predicts the omitted part based on the calculated parameters. Thus, Q^2 shows how well the data collection empirically can be reconstructed with the help of model and the PLS parameters. If

Table 5.11 Summary of structural model assessment

Validity type	Criterion	Results
Model validity	Coefficient of determination (R^2)	Values are 0.333 (INT), 0.419 (PM), 0.432 (SE), 0.545 (SECURITYCULTURE). Values obtained are considered moderate [26]
	Effect size (f^2)	Values indicate low to medium effect on INT [26]
	Predictive relevance (Q^2)	Values are 0.284 (INT), 0.312 (PM), 0.349 (SE), 0.456 (SECURITYCULTURE), which is higher than cut-off value of 0 indicating predictive relevance

the value of Q^2 is larger than 0, it implies that the framework has sufficient predictive relevance.

According to the results, as shown in Table 5.10, using an omission distance (D) of 7, this study obtains a Q^2 of intention to comply (0.284), perceived mandatoriness (0.312), self-efficacy (0.349) and security culture (0.456) are greater than 0, indicating the model's predictive relevance [27, 103] (Table 5.11).

5.3.14 *Discussion on Quantitative Findings*

This study empirically investigates the key factors affecting employees' intention to comply with an organization's BYOD security policies. It has added up some values to extant BYOD security policy compliance studies through the extension of constructs using integration of theories (OCT, SCT and security culture) to identify the factors that contribute to BYOD security policy compliance.

In answering RO3, the conceptual research framework is tested quantitatively by addressing nine (9) hypotheses. The nine hypotheses were developed to test the factors influencing employees' intention to comply with BYOD security policy. The convergence and discriminant validity test have been conducted before performing structural assessment. The summary of measurement model assessment is shown in Table 5.4 and research hypotheses testing results are summarized in Table 5.9.

The PLS-SEM path analysis revealed that, with the exception of security policy evaluation and security culture, the majority of the constructs in the BYOD security policy compliance is associated with employees' compliance behavior towards BYOD security policies. Seven out of nine hypotheses were supported, and contrary to the findings from past literature on security policy compliance, evaluation of security policy and security culture were found not significant, resulting in the rejection of hypotheses H2 and H8. Findings appeared to strongly support perceived mandatoriness, specification of security policy, BYOD IT support, SETA, self-efficacy and psychological ownership as strong predictors of security policy compliance intention. Figure 5.6 depicts a summary of the BYOD security policy compliance framework's statistical path results, where solid arrows indicate a significant relationship between

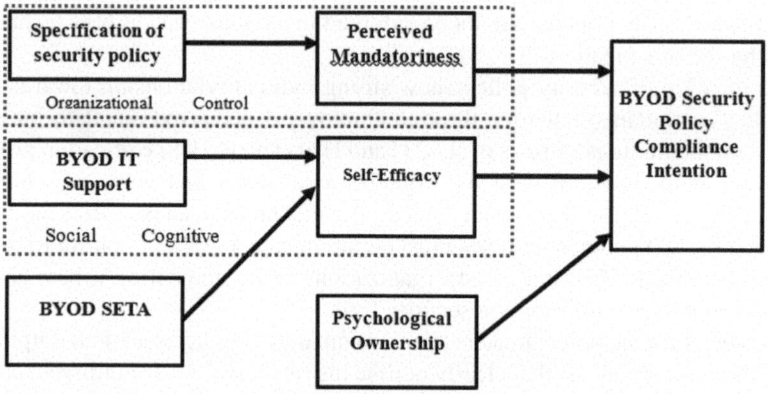

Fig. 5.7 Validated BYOD Security Policy Compliance Framework

the given components and dependent variable and dashed arrows indicate no effect. This fulfils RO3.

RO4 validates the model by assessing the R^2, f^2 and Q^2 to show the predictive relevance of the framework. Taken together, the framework's predictive accuracy (and the amount of variance in security policy compliance intention construct explained by all its exogenous constructs) was evaluated and the findings proved moderate predictive power ($R^2 = 0.333$) for security policy compliance intention. In addition, it was also found to possess moderate predictive relevance ($Q^2 = 0.284$). Therefore, RO4 was achieved.

The PLS-SEM path analysis found that the majority of the components in the BYOD security policy compliance model had an effect on employees' compliance behavior (Fig. 5.7). The only exceptions were the evaluation of security policies and security culture. The following section will discuss the hypotheses' findings.

5.3.15 *Outcomes of Perceived Mandatoriness, Specification of Security Policy and Evaluation of Security Policy*

The study finds that perceived mandatoriness play a key role in predicting BYOD security policy compliance, which is consistent with Boss et al. [21] and Galvez et al. [49]. Mandatoriness is critical for organizations to ensure that their critical information is secure and not compromised [45]. The results clearly show that employees understand the importance of securing information in the organization, which was the reason for mandatory security policy compliance requirement [101]. In most organizations, policies are the only security control measure to control employees, hence it is crucial to ensure BYOD policies are enforced effectively [121]. However, organizations have to keep in mind, BYOD is built on devices that organizations do not own, so security policies cannot be enforced in the same way as how technology owned by the organizations [67, 81].

Specification of security policy show strong indirect relationship towards security policy compliance intention through employees' perceived mandatoriness and aligns with the findings of Boss et al. [21] and Hsu et al. [63]. Specification provides employees with direction with the organizational goals and ways to achieve it. With the increased use of personal devices, organizations need prioritize the task of designing BYOD policies. Policies must communicate a sense of control to prevent IT security breach [75] and assist organizations in implementing robust security measures to improve information security [29].

It is crucial for the policy makers to communicate the policies all their employees. By disseminating policies that clearly outline the repercussions for failing to comply with security policy, employees will adhere to the policies [80]. Top management should articulate a clear vision and strategy for an effective information security policy and establish clear goals and objectives to attain security [13, 64]. Designing policies alone is not enough, clear and concise procedures on how to perform important security tasks, need to be provided [17, 121]. Overall, organizations should focus on establishing a comprehensive set of rules and procedures in a sufficiently clear and detailed manner in order to guide appropriate information security behavior [7].

Definitional confusion and contradiction lead to miscommunication and ineffective policies. Failure to specify the underlying assumption of the meaning of the term can lead to ambiguity [83]. Additionally, policies can highlight potential risks and protect organizational data from data leakage [67]. [58] discovered that there are concerns that must be addressed clearly, for example, employees and management must determine who owns the information on a personal device. Communicating a well-defined IS security policy that includes IS security goals directs individuals about where an organization is headed in terms of IS security.

Contrary to the hypotheses, the results indicate that evaluation of security policy is not significant towards perceived mandatoriness. It contradicted what was found by Cuganesan et al. [33], that evaluation procedures provide feedback on their behavior. One possible reason is that thorough evaluation consumes large amount of resources, manpower and time consuming [118].

5.3.16 Outcomes of Self-efficacy, BYOD IT Support and BYOD SETA

This study extended the concept of self-efficacy to empirically test its significance on security policy compliance intention and the association of variables (BYOD SETA and BYOD IT Support) on self-efficacy. Following previous studies, it is found that self-efficacy is positively associated with employees' security policy compliance intention. The relationship between self-efficacy and intention has not been consistent in prior research, hence this finding is useful. The significant effect implies that employees who perceive themselves having skills and capability to perform security tasks have the intention to comply with the security policy. It is critical that they have confidence in their ability to accomplish the necessary tasks as indicated in the security policy [2, 17, 22, 24, 41, 49, 61, 68, 82, 94, 109, 117].

As mobile device technology evolves, there is a need to continuously update and improve employee security capabilities [51], and the process should continue until employees embrace secure behavior [3]. Providing SETA to all levels of employees in organizations gives them the opportunity to impart general knowledge of IS security environment, along with the skills necessary to perform the required IS security tasks [109].

These findings support previous study [62]. Employees who see their employer's support will feel more inclined to take responsibility for the organization and its information resources [32]. Organizations have to allocate service desk resources and accommodate to employees' need with prominent areas of support for BYOD. Besides, receiving support such as this drives employees' commitment to comply to the security policy [34, 102]. Additionally, this is a more effective method of protecting corporate data than sending it to device service centers, where unscrupulous outsiders could get access to government systems and breach data, thereby exposing extremely sensitive information to foreign parties. Zahadat [121] recommended three approaches for providing employees with BYOD IT support. The first option is for the business to approve a vendor to offer all device maintenance and support, and organizations may then add controls and policies to protect the data on the devices. The second option for device maintenance is a hybrid plan, where the organization provides support for critical user problems while other issues are extended to the vendors. This option requires additional resources to set up the help desk and procedures to determine service scope and levels between the organization and vendors. The third option is for the organization to provide full support. Organization has the most control in this option, however there will be an increase of cost in training personnel and placing competent staffs to support various endpoints.

Training support staffs is important in order to assist employees with their devices so that they are safe and do not introduce malware or other threats into the organization network. To provide a good BYOD IT support, a clear demarcation must be outlined on the scope of support to avoid the IT help desk from being overwhelmed with the influx of devices.

BYOD SETA has been examined to see its effects on self-efficacy and results show SETA increases security policy compliance through self-efficacy as found in previous study [37, 120]. This is also in accordance with the requirements of the standard, which stipulate that enterprises must give the essential skills to enhance security management [65, 113]. As such, SETA programs should focus on providing employees with information security knowledge and skills to perform security tasks [120]. This is critical because personal devices pose additional security concerns compared to traditional computing environments. ISP compliance can be jeopardized by management's incorrect estimation of employees' competence to execute security procedures [113].

This suggests that employees build their self-efficacy by continuously acquiring information security expertise [41, 59] as extant research also cautioned that creating awareness is not a one-time effort [8]. Managers and leaders are responsible in ensuring that employees are provided with knowledge and strategies through thorough information security awareness [119].

SETA is shown to relate positively to security culture as suggested in [38, 86]. It proves to be the most effective way to cultivate security culture. Training programs need to be designed more effectively to foster self-efficacy belief [94]. All employees from the lowest rank to the highest-ranking officer need to attend SETA programs as each one of them have access to the organizational network and information [5]. From a newly joined employee to a senior employee, onboarding to routine refresher SETA should be the core of their culture to ensure all employees are aware and informed regarding organization's security practices to protect the confidentiality, integrity and availability of the organizational information [110].

SETA could be critical since it was found to enhance self-efficacy, security culture and security policy compliance intention. Organizations should invest in SETA programs to consequently create competent employees that will increase compliance intention as well as improve security culture in the organizations. Programs are needed to educate these employees on the latest trends in cyber-attacks and how to protect their devices. This helps in protecting the intellectual capital of the organizations [59].

SETA practices must go beyond annual security training (computer based training) and adopt a broader approach in order to change attitudes, perceptions, routines and assumptions, as well as developing new skills [10, 11]. SETA implementation in organizations need drastic and transformative change in order to ensure efficiency.

5.3.17 Outcomes of Security Culture

Unexpectedly, contrary to what was hypothesized; the study finds that security culture failed to demonstrate significance towards BYOD security policy compliance. This contradicts with the findings of previous studies [35, 52]. Though it was recommended by the experts in the qualitative study, it contradicts with the employees. This indicates that employees do not perceive security culture as effective or impactful to have a significant effect on their compliance intention towards the BYOD security policy. This is supported by Sommestad et al. [106] in their review that security culture is a weak predictor of intention to comply with information security policy. According to Schein [99, 100], the likely explanation for this non-significant link is that culture does not present itself easily. According to Vroom and Von Solms [118], an organization's culture could have either a positive or negative effect on information security. "Changing the perception or behavior of users towards a positive information security culture is not an easy or straightforward task" [97]. Additionally, building security culture in organizations requires tremendous effort and investment from all employees [10].

A strong culture exists when employees are aware of security risk, preventive measures and take responsibility. One possible explanation may be that culture of an organization is not formed by what management preaches or publishes, but what it accepts in practice. When senior management not only espouses that something is important, but also consistently behaves to support this message, employees will begin to believe what is espoused [48]. The findings may just reflect employees' perception of efforts made by management to attain information security compliance, based on their management support and involvement, goal alignment, and efficacy [72]. Top management that is sensitive to the need of information security will develop and enforce information security policy, which lead to the development of positive information security culture. It is only through such culture, security threats or attacks can be overcome or mitigated [77]. A study by Rajab and Eydgahi [92] also found that top management commitment is not significant towards information security policy compliance.

Some managements do view information security as an area belonging to the IT department, although the reality is that compliance begins with the top management to demonstrate the importance of it to all the employees [43]. If the commitment is perceived as creditable, employees will respond by trusting management and by making decisions in congruence with the championed initiatives, programs, or policies [66]. Additionally, lack of skills, knowledge and correct descriptions of information security terms, scope and task, pose challenges to top management from executing their responsibilities [20]. Compliance starts from the top level in the organization to reflect a positive image and then other employees will start to follow suit. Individual level motivating factors may play a larger role in the lack of a security culture, as information security is at the level of the person rather than being driven from the top down and across the enterprise [19].

BYOD presents less control over things due to ownership issues [16]. The lack of significance could also be attributed to the idea of monitoring devices which may make employees uneasy. Some organizations face challenges when it comes to monitoring or tracking their employees while still allowing them to operate in a flexible manner [15, 40]. Surveillance measure is a popular strategy in highly regulated industries as information privacy are of utmost concern and it has been found to impact employees security policy compliance intention previously [114]. Though monitoring systems can prove beneficial for detecting unusual activity or locating a lost device, many consider these solutions to venture too far into employees' privacy. Device monitoring raises internal misuse [90] and may burden personnel, leading to security policy non-compliance [104] or violate their privacy [74]. In the long run, employees may end up leaving the organization due to this privacy violation [67]. Tracking personal devices may also result in legal issues if not executed properly [121]. Technical security systems also exhaust and control employees during their job processes, according to studies [4, 36, 73].

It has also been suggested, that if a negative security culture exist, it could introduce risk, and organizations should identify interventions to align it with the positive security culture [39]. SETA is one intervention that has been identified through this study that can improve security culture of an organization.

5.3.18 Outcomes of Psychological Ownership

This study offers a contribution by utilizing the construct of psychological ownership as a significant predictor of employees to comply to BYOD security policies. This is similar with the findings of Thompson et al. [112]. The psychological state pushes employees towards security compliance. This finding has answered the call from McGill and Thompson [78], that the shift of applying psychology ownership has happened towards personal devices. This is relevant now that devices are used to store personal and financial information for many people and used for important transactions. Employees believe that if they do not follow the security policy, the security of their data and devices will be jeopardized, and hence feel obligated to comply [18].

5.4 Summary

Qualitative research approach was utilized to gather deeper insights about the research problem obtained from review of literature, along with establishing arguments for supporting the BYOD security policy compliance framework's exogenous constructs before being assessed and validated hypothetically. The results of risk assessment indicate 16 critical risks for public sector organizations and experts recommend Security Training and Awareness (SETA), Policy, Top Management Commitment and Technical Countermeasure to overcome the critical BYOD risks. These control measures were mapped into the conceptual framework to be tested in the following phase.

Moving on from the qualitative research, the framework was assessed using PLS-SEM statistical analysis. The first stage of analysis is to assess the reliability and validity. Using CFA, the framework fulfilled the reliability and validity evaluations. Hypotheses testing found that perceived mandatoriness, self-efficacy, psychological ownership as determinants of BYOD security policy compliance. Evaluation of security policy compliance behavior is found to be insignificant towards perceived mandatoriness. Security culture was found insignificant towards BYOD security policy compliance. This contradicts the findings from the literature [34]. Overall, the significant constructs provided the endogenous construct with predictive accuracy. Findings of both the qualitative and quantitative research are summarized and presented in next chapter. The qualitative and quantitative studies, when combined, address the research question by identifying factors that influence BYOD security policy compliance. The next chapter presents the overall contribution of this study.

References

1. Abdul Molok, N.N., Ahmad, A., Chang, S.: A case analysis of securing organisations against information leakage through online social networking. Int. J. Inf. Manag. **43**, 351–356 (2018). https://doi.org/10.1016/j.ijinfomgt.2018.08.013
2. Al-Emran, M., Granić, A., Al-Sharafi, M.A., Ameen, N., Sarrab, M.: Examining the roles of students' beliefs and security concerns for using smartwatches in higher education. J. Enterp. Inf. Manag. (2020). https://doi.org/10.1108/JEIM-02-2020-0052
3. Alassaf, M., Alkhalifah, A.: Exploring the influence of direct and indirect factors on information security policy compliance: a systematic literature review. IEEE Access **9**, 162687–162705 (2021). https://doi.org/10.1109/access.2021.3132574
4. Albrechtsen, E.: A qualitative study of users' view on information security. Comput. Secur. **26**(4), 276–289 (2007). https://doi.org/10.1016/j.cose.2006.11.004
5. AlGhamdi, S., Win, K.T., Vlahu-Gjorgievska, E.: Information security governance challenges and critical success factors: systematic review. Comput. Secur. **99**, 102030 (2020). https://doi.org/10.1016/j.cose.2020.102030
6. Alhogail, A.: Design and validation of information security culture framework. Comput. Hum. Behav. **49**, 567–575 (2015). https://doi.org/10.1016/j.chb.2015.03.054
7. AlKalbani, A., Deng, H., Kam, B.: Organisational Security Culture and information security compliance for E-Government development: the moderating effect of social pressure. In: 19th Pacific Asia Conference on Information Systems, PACIS 2015, p. 65 (2015). http://aisel.aisnet.org/pacis2015/65
8. Allam, S., Flowerday, S.V., Flowerday, E.: Smartphone information security awareness: a victim of operational pressures. Comput. Secur. **42**, 56–65 (2014). https://doi.org/10.1016/j.cose.2014.01.005
9. Alnatheer, M.A.: Information security culture critical success factors. In: Proceedings—12th International Conference on Information Technology: New Generations, ITNG 2015, pp. 731–735 (2015). https://doi.org/10.1109/ITNG.2015.124
10. Alshaikh, M.: Developing cybersecurity culture to influence employees behavior: a practice perspective. Comput. Secur. 102003 (2020). https://doi.org/10.1016/j.cose.2020.102003
11. Alshaikh, M., Maynard, S.B., Ahmad, A., Chang, S.: An exploratory study of current information security training and awareness practices in organizations. In: Proceedings of the 51st Hawaii International Conference on System Sciences, vol. 9, pp. 5085–5094 (2018). https://doi.org/10.24251/hicss.2018.635
12. Ameen, N., Tarhini, A., Hussain Shah, M., Madichie, N.O.: Employees' behavioural intention to smartphone security: a gender-based, cross-national study. Comput. Hum. Behav. **104**(106184) (2020). https://doi.org/10.1016/j.chb.2019.106184
13. Ameen, N., Tarhini, A., Shah, M.H., Madichie, N., Paul, J., Choudrie, J.: Keeping customers' data secure: a cross-cultural study of cybersecurity compliance among the Gen-Mobile workforce. Comput. Hum. Behav. **114** (2021). https://doi.org/10.1016/j.chb.2020.106531
14. Anderson, J., Gerbing, D.W.: Structural equation modeling in practice: a review and recommended two-step approach. Psychol. Bull. **103**(3), 411–423 (1988). https://doi.org/10.1037/0033-2909.103.3.411
15. Bain, P., Taylor, P.: Entrapped by the "electronic panopticon"? Worker resistance in the call centre. N. Technol. Work. Employ. **15**(1), 2–18 (2000). https://doi.org/10.1111/1468-005X.00061
16. Barlette, Y., Jaouen, A., Baillette, P.: Bring Your Own Device (BYOD) as reversed IT adoption: Insights into managers' coping strategies. Int. J. Inf. Manag. **56**(102212) (2021). https://doi.org/10.1016/j.ijinfomgt.2020.102212
17. Belanger, F., Crossler, R.E.: Dealing with digital traces: understanding protective behaviors on mobile devices. J. Strateg. Inf. Syst. **28**(1), 34–49 (2018). https://doi.org/10.1016/j.jsis.2018.11.002

18. Blythe, J.M., Coventry, L.: Costly but effective: comparing the factors that influence employee anti-malware behaviours. Comput. Hum. Behav. **87**, 87–97 (2018). https://doi.org/10.1016/j.chb.2018.05.023
19. Blythe, J.M., Coventry, L., Little, L.: Unpacking security policy compliance: the motivators and barriers of employees' security behaviors. In: SOUPS 2015—Proceedings of the 11th Symposium on Usable Privacy and Security, pp. 103–122 (2019)
20. Bongiovanni, I., Renaud, K., Brydon, H., Blignaut, R., Cavallo, A.: A quantification mechanism for assessing adherence to information security governance guidelines. Inf. Comput. Secur. (2022) (ahead-of-print). https://doi.org/10.1108/ics-08-2021-0112
21. Boss, S.R., Kirsch, L.J., Angermeier, I., Shingler, R.A., Boss, R.W.: If someone is watching, I'll do what I'm asked: mandatoriness, control, and information security. Eur. J. Inf. Syst. **18**(2), 151–164 (2009). https://doi.org/10.1057/ejis.2009.8
22. Bulgurcu, B., Cavusoglu, H., Benbasat, I.: Information security policy compliance: an empirical study of rationality-based beliefs and information security awareness. MIS Q. **34**(3), 523–548 (2010). https://doi.org/10.1093/bja/aeq366
23. Cain, M.K., Zhang, Z., Yuan, K.H.: Univariate and multivariate skewness and kurtosis for measuring nonnormality: prevalence, influence and estimation. Behav. Res. Methods **49**(5), 1716–1735 (2017). https://doi.org/10.3758/s13428-016-0814-1
24. Chan, M., Woon, I., Kankanhalli, A.: Perceptions of information security in the workplace: linking information security climate to compliant behavior. J. Inf. Privacy Secur. **1**(3), 18–41 (2005). https://doi.org/10.1080/15536548.2005.10855772
25. Chen, H., Li, Y., Chen, L., Yin, J.: Understanding employees' adoption of the Bring-Your-Own-Device (BYOD): the roles of information security-related conflict and fatigue. J. Enterp. Inf. Manag. (2020). https://doi.org/10.1108/JEIM-10-2019-0318
26. Chin, W.W.: The partial least squares approach for structural equation modeling. Modern Methods for Business Research, pp. 295–336 (1998)
27. Chin, W.W., Cheah, J.H., Liu, Y., Ting, H., Lim, X.J., Cham, T.H.: Demystifying the role of causal-predictive modeling using partial least squares structural equation modeling in information systems research. Ind. Manag. Data Syst. **120**(12), 2161–2209 (2020). https://doi.org/10.1108/IMDS-10-2019-0529
28. Chin, W.W., Marcolin, B.L., Newsted, P.R.: A partial least squares latent variable modeling approach for measuring interaction effects: results from a Monte Carlo simulation study and an electronic-mail emotion/adoption study. Inf. Syst. Res. **14**(2) (2003). https://doi.org/10.1126/science.143.3610.994
29. Chua, H.N., Teh, J.S., Herbland, A.: Identifying the effect of data breach publicity on information security awareness using hierarchical regression. IEEE Access PP, 1–1 (2021). https://doi.org/10.1109/access.2021.3107426
30. Cohen, J.: Statistical power analysis for the behavioural sciences. Lawrence Erlbaum (1988)
31. Cram, W.A., Proudfoot, J.G., D'Arcy, J.: Maximizing employee compliance with cybersecurity policies. MIS Q. Exec. **19**(3), 183–198 (2020). https://doi.org/10.17705/2msqe.00032
32. Crossler, R.E., Ogbanufe, O., Biros, D.: Exploring stewardship: a precursor to voluntary security behaviors. Comput. Secur. 102397 (2021). https://doi.org/10.1016/j.cose.2021.102397
33. Cuganesan, S., Steele, C., Hart, A.: How senior management and workplace norms influence information security attitudes and self-efficacy. Behav. Inf. Technol. **37**(1), 50–65 (2018). https://doi.org/10.1080/0144929X.2017.1397193
34. D'Arcy, J., Greene, G.: Security culture and the employment relationship as drivers of employees' security compliance. Inf. Manag. Comput. Secur. **22**(5), 474–489 (2014). https://doi.org/10.1108/IMCS-08-2013-0057
35. D'Arcy, J., Greene, G.: The multifaceted nature of security culture and its influence on end user behavior. In: IFIP TC 8 International Workshop on Information Systems Security Research, pp. 145–157 (2009)

36. D'Arcy, J., Herath, T., Shoss, M.K.: Understanding employee responses to stressful information security requirements: a coping perspective. J. Manag. Inf. Syst. **31**(2), 285–318 (2014). https://doi.org/10.2753/MIS0742-1222310210
37. D'Arcy, J., Hovav, A., Galletta, D.: User awareness of security countermeasures and its impact on information systems misuse: a deterrence approach. Inf. Syst. Res. **20**(1), 79–98 (2009). https://doi.org/10.1287/isre.1070.0160
38. da Veiga, A., Astakhova, L.V., Botha, A., Herselman, M.: Defining organisational information security culture—perspectives from academia and industry. Comput. Secur. **92**, 101713 (2020). https://doi.org/10.1016/j.cose.2020.101713
39. da Veiga, A., Martins, N.: Defining and identifying dominant information security cultures and subcultures. Comput. Secur. **70**, 72–94 (2017). https://doi.org/10.1016/j.cose.2017.05.002
40. de Vargas Pinto, A., Maçada, A.C.G., Mallmann, G.L.: Workaround behaviour in information systems research. Revista de Gestão **25**(4), 430–446 (2018). https://doi.org/10.1108/rege-12-2017-0015
41. Dhillon, G., Talib, Y.Y.A., Picoto, W.N.: The mediating role of psychological empowerment in information security compliance intentions. J. Assoc. Inf. Syst. **21**(1), 152–174 (2020). https://doi.org/10.17705/1jais.00595
42. Diamantopoulos, A., Siguaw, J.A.: Formative versus reflective indicators in organizational measure development: a comparison and empirical illustration. Br. J. Manag. **17**(4), 263–282 (2006). https://doi.org/10.1111/j.1467-8551.2006.00500.x
43. Dzazali, S., Sulaiman, A., Zolait, A.H.: Information security landscape and maturity level: case study of Malaysian Public Service (MPS) organizations. Gov. Inf. Q. **26**(4), 584–593 (2009). https://doi.org/10.1016/j.giq.2009.04.004
44. Evans, M., He, Y., Maglaras, L., Yevseyeva, I., Janicke, H.: Evaluating information security core human error causes (IS-CHEC) technique in public sector and comparison with the private sector. Int. J. Med. Informatics **127**, 109–119 (2019). https://doi.org/10.1016/j.ijmedinf.2019.04.019
45. Farshadkhah, S., Van Slyke, C., Fuller, B.: Onlooker effect and affective responses in information security violation mitigation. Comput. Secur. **100**, 102082 (2021). https://doi.org/10.1016/j.cose.2020.102082
46. Fornell, C., Larcker, D.F.: Evaluating structural equation models with unobservable variables and measurement errors. J. Mark. Res. **18**(1), 39–50 (1981)
47. Franke, G., Sarstedt, M.: Heuristics versus statistics in discriminant validity testing: a comparison of four procedures. Internet Res. **29**(3), 430–447 (2019). https://doi.org/10.1108/IntR-12-2017-0515
48. Furnell, S., Thomson, K.L.: From culture to disobedience: recognising the varying user acceptance of IT security. Comput. Fraud Secur. **2009**(2), 5–10 (2009). https://doi.org/10.1016/S1361-3723(09)70019-3
49. Galvez, S.M., Shackman, J.D., Guzman, I.R., Ho, S.M.: Factors affecting individual information security practices. In: SIGMIS-CPR'15, June, vol. 26(8), 555–557 (2015). https://doi.org/10.1145/2751957.2751966
50. Gillam, A.R., Foster, W.T.: Factors affecting risky cybersecurity behaviors by U.S. workers: an exploratory study. Comput. Hum. Behav. **108**(106319) (2020). https://doi.org/10.1016/j.chb.2020.106319
51. Giwah, A.D.: Empirical Assessment of Mobile Device Users' Information Security Behavior towards Data Breach: Leveraging Protection Motivation Theory (2019)
52. Greene, G., D'Arcy, J.: Assessing the impact of security culture and the employee-organization relationship in IS security compliance. In: 5th Annual Symposium on Information Assurance (ASIA'10), pp. 42–49 (2010). https://doi.org/10.1108/IMCS-08-2013-0057
53. Guri, M., Puzis, R., Choo, K.-K.R., Rubinshtein, S., Kedma, G., Elovici, Y.: Using malware for the greater good: mitigating data leakage. J. Netw. Comput. Appl. **145**, 102405 (2019). https://doi.org/10.1016/j.jnca.2019.07.006
54. Hair, J.F., Hult, G.T.M., Ringle, C.M., Sarstedt, M., Thiele, K.O.: Mirror, mirror on the wall: a comparative evaluation of composite-based structural equation modeling methods. J. Acad. Mark. Sci. **45**(5), 616–632 (2017). https://doi.org/10.1007/s11747-017-0517-x

55. Hair, J.F., Hult, T.M., Ringle, C.M., Sarstedt, M.: A primer on partial least squares structural equation modeling (PLS-SEM). Sage Publications (2017). https://doi.org/10.1080/1743727x.2015.1005806
56. Hair, J.F., Ringle, C.M., Sarstedt, M.: PLS-SEM: indeed a silver bullet. J. Mark. Theory Pract. **19**(2), 139–152 (2011). https://doi.org/10.2753/MTP1069-6679190202
57. Hair, J.J.F., Matthews, L.M., Matthews, R.L., Sarstedt, M.: PLS-SEM or CB-SEM: updated guidelines on which method to use. Int. J. Multivar. Data Anal. **1**(2), 107 (2017). https://doi.org/10.1504/ijmda.2017.087624
58. Harris, J., Ives, L., Junglas, I.: IT consumerization: when gadgets turn into enterprise IT tools. MIS Q. Exec. **11**(3), 99–112 (2012)
59. He, W., Ash, I., Anwar, M., Li, L., Yuan, X., Xu, L., Tian, X.: Improving employees' intellectual capacity for cybersecurity through evidence-based malware training. J. Intellect. Cap. **21**(2), 203–213 (2020). https://doi.org/10.1108/JIC-05-2019-0112
60. Henseler, J., Ringle, C.M., Sarstedt, M.: A new criterion for assessing discriminant validity in variance-based structural equation modeling. J. Acad. Mark. Sci. **43**(1), 115–135 (2015). https://doi.org/10.1007/s11747-014-0403-8
61. Herath, T., Rao, H.R.: Protection motivation and deterrence: a framework for security policy compliance in organisations. Eur. J. Inf. Syst. **18**(2), 106–125 (2009). https://doi.org/10.1057/ejis.2009.6
62. Hovav, A., Putri, F.F.: This is my device! Why should I follow your rules? Employees' compliance with BYOD security policy. Pervasive Mob. Comput. **32**, 35–49 (2016). https://doi.org/10.1016/j.pmcj.2016.06.007
63. Hsu, J.S.C., Shih, S.-P., Wen Hung, Y., Benjamin Lowry, P.: The role of extra-role behaviors and social controls in information security policy effectiveness. Inf. Syst. Res. **26**(2), 0–19 (2015)
64. Hu, Q., Dinev, T., Hart, P., Cooke, D.: Managing employee compliance with information security policies: the critical role of top management and organizational culture. Decis. Sci. **43**(4), 615–660 (2012). https://doi.org/10.1111/j.1540-5915.2012.00361.x
65. International Organization for Standardization: ISO/IEC 27001:2013—Information technology—Security techniques—Information security management systems—Requirements. ISO (2013)
66. James, H.S.: Reinforcing ethical decision making through organizational structure. J. Bus. Ethics **28**, 43–58 (2000)
67. Jarrahi, M.H., Crowston, K., Bondar, K., Katzy, B.: A pragmatic approach to managing enterprise IT infrastructures in the era of consumerization and individualization of IT. Int. J. Inf. Manag. **37**(6), 566–575 (2017). https://doi.org/10.1016/j.ijinfomgt.2017.05.016
68. Johnston, A.C., Warkentin, M.: Fear appeal and information security behaviour: an empirical study. MIS Q. **34**(1), 1–20 (2010)
69. Jones, B.H., Chin, A.G.: On the efficacy of smartphone security: a critical analysis of modifications in business students' practices over time. Int. J. Inf. Manag. **35**(5), 561–571 (2015). https://doi.org/10.1016/j.ijinfomgt.2015.06.003
70. Keen, R: Actions to Address Risks and Opportunities Explained [with procedure]. ISO 9001 Checklist (2021). https://www.iso-9001-checklist.co.uk/6.1-actions-to-address-risks-and-opportunities.htm
71. Kirsch, L.J.: Deploying common systems globally: the dynamics of control. Inf. Syst. Res. **15**(4), 374–395 (2004). https://doi.org/10.1287/isre.1040.0036
72. Knapp, K.J., Marshall, T.E., Rainer, R.K., Ford, F.N.: Information security: management's effect on culture and policy. Inf. Manag. Comput. Secur. **14**(1), 24–36 (2006). https://doi.org/10.1108/09685220610648355
73. Lee, C., Lee, C.C., Kim, S.: Understanding information security stress: focusing on the type of information security compliance activity. Comput. Secur. **59**, 60–70 (2016). https://doi.org/10.1016/j.cose.2016.02.004
74. Lee, J., Crossler, R.E., Warkentin, M.: Implications of monitoring mechanisms on Bring Your Own Device (BYOD) adoption. In: 34th International Conference on Information Systems (ICIS 2013), pp. 1–12 (2013)

75. Lowry, P.B., Teh, N., Molyneux, B., Bui, S.N.: Be careful what you wish for: proposing the IT policy control- reactance model (ITPCRM) to predict professional's intent to comply with new IT security policies along with their resulting anger. In: MIS Quarterly Pre-ICIS Workshop for Authors at the International Conference on System Sciences, pp. 1–40 (2010)
76. Ma, X.: IS professionals' information security behaviors in Chinese IT organizations for information security protection. Inf. Process. Manag. **59**(1), 102744 (2022). https://doi.org/10.1016/j.ipm.2021.102744
77. Masrek, M.N., Harun, Q.N., Ramli, I.: The role of top management in information security practices. In: 6th International Conference on Education, Social Sciences and Humanities (SOCIOINT 2019), pp. 983–990 (2019)
78. McGill, T., Thompson, N.: Old risks, new challenges: exploring differences in security between home computer and mobile device use. Behav. Inf. Technol. **36**(11), 1111–1124 (2017). https://doi.org/10.1080/0144929X.2017.1352028
79. McNeish, D.: Thanks coefficient alpha, We'll take it from here. Psychol. Methods **23**(3), 412–433 (2018). https://doi.org/10.1037/met0000144
80. Merhi, M.I., Ahluwalia, P.: Examining the impact of deterrence factors and norms on resistance to Information Systems Security. Comput. Hum. Behav. **92**, 37–46 (2019). https://doi.org/10.1016/j.chb.2018.10.031
81. Miller, K.W., Voas, J., Hurlburt, G.F.: BYOD: security and privacy considerations. IT Prof. **14**(5), 53–55 (2012). https://doi.org/10.1109/MITP.2012.93
82. Ng, B.Y., Kankanhalli, A., Xu, Y. (Calvin): Studying users' computer security behavior: a health belief perspective. Decis. Support Syst. **46**(4), 815–825 (2009).https://doi.org/10.1016/j.dss.2008.11.010
83. Paananen, H., Lapke, M., Siponen, M.: State of the art in information security policy development. Comput. Secur. **88**, 101608 (2020). https://doi.org/10.1016/j.cose.2019.101608
84. Palanisamy, R., Norman, A.A., Kiah, M.L.M.: BYOD security risks and mitigation strategies: insights from IT security experts. J. Organ. Comput. Electron. Commerc. **31**(4), 1–23 (2022). https://doi.org/10.1080/10919392.2022.2028530
85. Palanisamy, R., Norman, A.A., Kiah, M.L.M.: Compliance with Bring Your Own Device security policies in organizations: a systematic literature review. Comput. Secur. 101998 (2020). https://doi.org/10.1016/j.cose.2020.101998
86. Parsons, K.M., Young, E., Butavicius, M.A., McCormac, A., Pattinson, M.R., Jerram, C.: The influence of organizational information security culture on information security decision making. J. Cogn. Eng. Decis. Mak. **9**(2), 117–129 (2015). https://doi.org/10.1177/1555343415575152
87. Pienta, D., Thatcher, J.B., Johnston, A.C.: Protecting a whale in a sea of phish. J. Inf. Technol. (2020). https://doi.org/10.1177/0268396220918594
88. Podsakoff, P.M., MacKenzie, S.B., Lee, J.Y., Podsakoff, N.P.: Common method biases in behavioral research: a critical review of the literature and recommended remedies. J. Appl. Psychol. **88**(5), 879–903 (2003). https://doi.org/10.1037/0021-9010.88.5.879
89. Podsakoff, P.M., MacKenzie, S.B., Podsakoff, N.P.: Recommendations for creating better concept definitions in the organizational, behavioral, and social sciences. Organ. Res. Methods **19**(2), 159–203 (2016). https://doi.org/10.1177/1094428115624965
90. Posey, C., Bennett, B., Roberts, T., Lowry, P.: When computer monitoring backfires: invasion of privacy and organizational injustice as precursors to computer abuse. J. Inf. Syst. Secur. **7**(1), 24–47 (2011)
91. Posey, C., Roberts, T.L., Lowry, P.B., Hightower, R.T.: Bridging the divide: a qualitative comparison of information security thought patterns between information security professionals and ordinary organizational insiders. Inf. Manag. **51**(5), 551–567 (2014). https://doi.org/10.1016/j.im.2014.03.009
92. Rajab, M., Eydgahi, A.: Evaluating the explanatory power of theoretical frameworks on intention to comply with information security policies in higher education. Comput. Secur. **80**, 211–223 (2019). https://doi.org/10.1016/j.cose.2018.09.016

93. Ramayah, T., Cheah, J.-H., Chuah, F., Ting, H., Memon, M.A.: Partial Least Squares Structural Equation Modeling (PLS-SEM) using SmartPLS 3.0. Handbook of Market Research, 2nd edn. Pearson Malaysia Sdn Bhd. (2018)
94. Rhee, H.S., Kim, C., Ryu, Y.U.: Self-efficacy in information security: Its influence on end users' information security practice behavior. Comput. Secur. **28**(8), 816–826 (2009). https://doi.org/10.1016/j.cose.2009.05.008
95. Richardson, H.A., Simmering, M.J., Sturman, M.C.: A tale of three perspectives: examining Post Hoc statistical techniques for detection and correction of Common Method Variance. Organ. Res. Methods **12**(4) (2009)
96. Ringle, C.M., Wende, S., Becker, J.M.: SmartPLS 3.0 (2015). www.smartpls.com
97. Safa, N.S., Sookhak, M., Von Solms, R., Furnell, S., Ghani, N.A., Herawan, T.: Information security conscious care behaviour formation in organizations. Comput. Secur. **53**, 65–78 (2015). https://doi.org/10.1016/j.cose.2015.05.012
98. Sanyal, P., Menon, N., Siponen, M.: An empirical examination of the economics of mobile application security. MIS Q. **45**(4), 2235–2260 (2021). https://doi.org/10.25300/MISQ/2021/15315
99. Schein, E.H.: Organizational culture and leadership. Jossey Bass, San Fransissco (1985). https://doi.org/10.1016/0024-6301(93)90120-5
100. Schein, E.H.: Organizational Culture and Leadership, 3rd edn. Jossey-Bass (2004)
101. Shao, X., Siponen, M., Liu, F.: Shall we follow? Impact of reputation concern on information security managers' investment decisions. Comput. Secur. **97**, 101961 (2020). https://doi.org/10.1016/j.cose.2020.101961
102. Sharma, S., Warkentin, M.: Do I really belong?: Impact of employment status on information security policy compliance. Comput. Secur. **87**, 101397 (2019). https://doi.org/10.1016/j.cose.2018.09.005
103. Shmueli, G., Sarstedt, M., Hair, J.F., Cheah, J.H., Ting, H., Vaithilingam, S., Ringle, C.M.: Predictive model assessment in PLS-SEM: guidelines for using PLSpredict. Eur. J. Mark. **53**(11), 2322–2347 (2019). https://doi.org/10.1108/EJM-02-2019-0189
104. Singh, T., Johnston, A.C., Thatcher, J.B.: How much is too much: employee monitoring, surveillance, and strain. In: Fortieth International Conference on Information Systems, pp. 1–9 (2019)
105. Smith, W.P.: Can we borrow your phone? Employee privacy in the BYOD era. J. Inf. Commun. Ethics Soc. **15**(4), 397–411 (2017). https://doi.org/10.1108/JICES-09-2015-0027
106. Sommestad, T., Hallberg, J., Lundholm, K., Bengtsson, J.: Variables influencing information security policy compliance: a systematic review of quantitative studies. Inf. Manag. Comput. Secur. **22**(1), 42–75 (2014). https://doi.org/10.1108/IMCS-08-2012-0045
107. Steelman, Z.R., Lacity, M., Sabherwal, R.: Charting your organization's Bring-Your-Own-Device voyage. MIS Q. **15**(2), 99–112 (2016)
108. Sullivan, G.M., Feinn, R.: Using effect size—or why the p value is not enough. J. Grad. Med. Educ. **4**(3), 279–282 (2012). https://doi.org/10.4300/jgme-d-12-00156.1
109. Talib, Y.Y.A.: Intrinsic motivation and Information Systems security policy compliance in organizations (2015)
110. Tam, C., De Matos, C., Oliveira, T.: What influences employees to follow security policies ? Saf. Sci. **147** (2022). https://doi.org/10.1016/j.ssci.2021.105595
111. Tehseen, S., Ramayah, T., Sajilan, S.: Testing and controlling for common method variance: a review of available methods. J. Manag. Sci. **4**(2), 142–168 (2017). https://doi.org/10.20547/jms.2014.1704202
112. Thompson, N., McGill, T.J., Wang, X.: "Security begins at home": determinants of home computer and mobile device security behavior. Comput. Secur. **70**, 376–391 (2017). https://doi.org/10.1016/j.cose.2017.07.003
113. Tsohou, A., Holtkamp, P.: Are users competent to comply with information security policies? An analysis of professional competence models. Inf. Technol. People **31**(5), 1047–1068 (2018). https://doi.org/10.1108/ITP-02-2017-0052

114. Tu, Z., Adkins, J., Zhao, G.Y.: Complying with BYOD security policies: a moderation model based on Protection Motivation Theory. J. Midwest Assoc. Inf. Syst. **2019**(1), 11–28 (2019). https://doi.org/10.17705/3jmwa.000045
115. Tu, Z., Turel, O., Yuan, Y., Archer, N.: Learning to cope with information security risks regarding mobile device loss or theft: an empirical examination. Inf. Manag. **52**(4), 506–517 (2015). https://doi.org/10.1016/j.im.2015.03.002
116. Urbach, N., Ahlemann, F.: Structural equation modeling in information systems research using partial least squares. J. Inf. Technol. Theory Appl. **11**(2), 5–40 (2010)
117. Verkijika, S.F.: Understanding smartphone security behaviors: an extension of the protection motivation theory with anticipated regret. Comput. Secur. **77**, 860–870 (2018). https://doi.org/10.1016/j.cose.2018.03.008
118. Vroom, C., Von Solms, R.: Towards information security behavioural compliance. Comput. Secur. **23**(3), 191–198 (2004). https://doi.org/10.1016/j.cose.2004.01.012
119. White, T., Lao, T.: IT managers' and IT professionals' mobile device security strategies: a qualitative exploratorary case study. J. Manag. Policy Pract. **22**(3), 90–100 (2020). https://doi.org/10.33423/jmpp.v22i3.4684
120. Yoo, C.W., Sanders, G.L., Cerveny, R.P.: Exploring the influence of flow and psychological ownership on security education, training and awareness effectiveness and security compliance. Decis. Support Syst. **108**, 107–118 (2018). https://doi.org/10.1016/j.dss.2018.02.009
121. Zahadat, N.: Mobile Security: A Systems Engineering Framework for Implementing Bring Your Own Device (BYOD) Security Through the Combination of Policy Management and Technology. ProQuest Dissertations and Theses, pp. 1–202 (2016). http://search.proquest.com.library.capella.edu/docview/1769079590?accountid=27965%5Cnhttp://wv9lq5ld3p.search.serialssolutions.com.library.capella.edu/?ctx_ver=Z39.88-2004&ctx_enc=info:ofi/enc:UTF-8&rfr_id=info:sid/ProQuest+Dissertations+&+Theses+Global&r
122. Zang, W.L.: Research of information security quantitative evaluation method. Appl. Mech. Mater. **513–517**, 369–372 (2014). https://doi.org/10.4028/www.scientific.net/AMM.513-517.369

Chapter 6
Conclusion and Future Works

Previous studies on BYOD security policy compliance had informed us that little research has been accomplished on identifying factors that influence security policy compliance amongst CNII agencies' employees. Drawing from various theories namely, Organizational Control Theory, Social Cognitive Theory, security culture, additional factors such as SETA and psychological ownership, this study fills the gap and contributes to the body of knowledge and policy to improve employees' BYOD security policy compliance.

This chapter discusses the research overview, reflection on how the research objectives were achieved, theoretical, practical, and methodological contributions, as well as the limitations. Suggestions for future research and development are also given. Finally, the conclusion ends this chapter.

6.1 Achievement of Research Objectives

This research was undertaken to investigate factors influencing BYOD security policy compliance in the Malaysian public sector. It began by looking at the BYOD practices, the benefits BYOD offers [4] and the security risks that it presents. In order to strengthen the BYOD security, based on this backdrop, the study produced a validated BYOD security policy compliance framework by examining the antecedents of BYOD security policy compliance intention.

This study attempted to fill in the gaps found on literature, where the study on BYOD is still scarce, and among the existing studies, almost none focused on the public sector. The majority of research concentrated on traditional computing security policy compliance, excluding the BYOD context. Little is known about compliance factors identification based on BYOD security risks or deploying other theories besides PMT [9] as most of the extant studies adopted existing frameworks [2]. People being the weakest link, cause security risk to the organization due to non-compliance

© The Author(s), under exclusive license to Springer Nature Switzerland AG 2025
R. Palanisamy et al., *Bring Your Own Device Security Policy Compliance Framework*, Information Systems Engineering and Management 37,
https://doi.org/10.1007/978-3-031-86801-6_6

behavior. Identifying compliance factors based on risks mitigating control measures is crucial as it helps the management to decide and execute the right strategies to cope with risk that affects the organization the most. Research has found that the motivation for security is to mitigate risk and reduce risk likelihood [8]. Hence, the study was designed using exploratory sequential design mixed method approach with the aim to develop a validated framework of BYOD security policy compliance factors [7].

The research objectives (RO) of this study are as follows:

a. To explore factors influencing BYOD security policy compliance intention.
b. To identify the risks caused by employees' non-compliance behavior towards BYOD security policy and control measures to mitigate the risk.
c. To develop the BYOD security policy compliance framework.
d. To validate the BYOD Security Policy Compliance Framework.

The following section elaborates as to how objectives were achieved through this study.

6.1.1 RO1: To Explore Factors Influencing BYOD Security Policy Compliance Intention

The activities to achieve RO1 started from the very beginning of the study, through performing a systematic literature review. In Phase 1, to address RO1, which is to establish the research framework, all studies in the area of BYOD security policy compliance were reviewed thoroughly by performing systematic reviews. Although security policy compliance factors have been widely researched [3], their relevance to the BYOD context is under researched. The review provided an overview of the theories used to explain and predict security behaviors in this field and a detailed understanding of factors that influence BYOD security policy compliance behavior. Among the findings raised were the lack of other theories employed besides PMT and BYOD IT support as an important factor that needs to be given attention by organizations as there will be more employee-owned devices used in the workplace than organization-owned devices. Lack of emphasis on BYOD security policy and security culture. Despite the importance of security culture, empirical research in this field remains scarce [14]. The plausible BYOD security policy compliance factors identified through the systematic review were interrelated and conceptualized to assess their significance towards intentions to comply with BYOD security policy in the later phase.

6.1.2 RO2: To Identify the Risks Caused by Employees' Non-compliance Behavior Towards BYOD Security Policy and Control Measures to Mitigate the Risk

Next, in Phase 2, to answer RO2, the study reviewed BYOD security risk systematically and explored the potential BYOD security policy compliance factors through an exploratory study by conducting interviews and risk assessment. This research question aimed to understand what motivates and prevents security risk in organizations by identifying risk reduction strategies. A total of eight experts representing five security organizations were interviewed to obtain their insights on BYOD practice and security risk encountered. Experts perform risks assessment to identify critical risk and recommend control measures to overcome them. As a result, 16 critical risks were identified and experts recommend control measures such as Policy, SETA, Top Management and Technical Countermeasures to overcome them. The proposed control measures were translated into constructs and conceptualized to form a research framework to be examined for their influence on BYOD security policy compliance intention. Identifying compliance factors based on control measures is established as an approach to address BYOD security risk while motivating employees to adopt compliance behavior as prescribed in the BYOD security policy. Thus, the RO2 was achieved.

6.1.3 RO3: To Develop the BYOD Security Policy Compliance Framework

In Phase 3, RO3 was realized by a quantitative method which was applied to investigate the influence of the proposed factors on BYOD security policy compliance intention. The factors were investigated by collecting data through a survey from 346 participants in three Critical National Information Infrastructure (CNII) agencies. The findings confirmed that perceived mandatoriness, self-efficacy and psychological ownership play a key role in determining employees' security policy compliance intention. Analysis also showed that specification of policy has a significant relationship with perceived mandatoriness. The study also found SETA and BYOD IT support demonstrating significance towards self-efficacy. Evaluation of compliance behavior was shown to be insignificant towards perceived mandatoriness. Analysis also revealed that security culture is not significantly associated with security policy compliance intention. However, this can be improved by implementing good SETA strategies as the study found SETA to be directly associated with security culture.

6.1.4 *RO4: To Validate the BYOD Security Policy Compliance Framework*

In the similar phase, RO4 was accomplished by performing coefficient of Determination, Level of effect size and Predictive Relevance using PLS-SEM. Results indicate that overall, the framework can predict up to 33.3% of the factors influencing BYOD Security Policy Compliance Intention, which is satisfactory. Additionally, looking at the effect size (f^2) values, it can be observed that all the relationship possesses substantive impact on the outcome at various degrees. Finally, PLS-Predict was performed and results indicate the model's predictive relevance.

6.2 Research Contributions

The findings from this study deepens the understanding on the factors that influence BYOD security policy compliance in the Malaysian public sector. The following sections presents the theoretical, practical, and methodological contributions of the study.

6.2.1 *Theoretical Contributions*

This research offers a few contributions to the existing IS literature. Numerous studies indicate that the cause for majority of the data breaches in the organizations is due to employees' non-compliance behavior towards security policy. Accordingly, this research investigates the factors influencing BYOD security policy compliance among Malaysian public sector employees via integration of theories (Organizational Control Theory, Social Cognitive Theory and Security Culture).

Identification of BYOD security risks and control measures to mitigate security risks

The study has reviewed BYOD security risk caused from security policy non-compliance and mapped the identified risks according to the People, Process and Technology (PPT) Model. Results obtained from the review study establish a clear understanding on BYOD features, security policy compliance theories and security policy compliance factors. Interviews and risk assessment resulted BYOD security risks according to the risk severity with recommended control measures to overcome the risks. Performing security risk assessment paves the way for future research to incorporate factors that encourage security policy compliance behavior. Policy, SETA, top management commitment and technical countermeasures were factors recommended by experts, which were interpreted theoretically into Organizational

Control Theory and security culture to provide a basis for BYOD security policy framework.

Integration of theories (Social Cognitive Theory, Control Theory) and under-explored factors to form research framework

The framework is based on the combination of two theories, quite rarely used in the area of information security; SCT and OCT, with security culture. This approach moves away from the PMT framework to a relatively new context and by adopting a mixed-methods approach it has provided better clarity of the use of theories. The study sheds light on an unexplored area of research of using different theories to explore compliance factors among public sector employees. In this study, the plausible factors were gathered from qualitative study and systematic reviews, and later empirically validated the causal links of employees' BYOD security policy compliance framework. By including the views of both security experts and the employees in both stages, this research broadens knowledge regarding BYOD ISP compliance from the view of both cohorts. Linking factors obtained from risk assessment to security policy compliance, the study provides a more nuanced understanding of the behavioral aspect of information security management.

Results yielded contribution to the concept of mandatoriness in the context of BYOD security policy compliance. The results of this study show that when employees' view a policy to be mandatory, they will comply and not according to volitional milieu in most cases in organizations. Specification of policy is indirectly associated with security policy compliance suggesting that codifying required behavior contributes to the perception of mandatoriness. Results do not support the significance of evaluation of compliance towards perceived mandatoriness. The insignificant relationship of evaluation, may indicate that employees are unenthusiastic towards being evaluated by their superiors for compliance behavior.

This research advances the existing body of knowledge by applying SCT in predicting the association between IT Support and self-efficacy towards employees' BYOD security policy compliance. Although SCT has been widely used in different fields, almost none was used to study the context of BYOD security policy compliance. The findings yielded a clear finding that BYOD IT support is indirectly associated with security policy compliance. In addition, results obtained suggest that employees find security training and awareness to be important. If organization can provide effective security training, employees' skills in using security tools will improve. The greater confidence employees have in their abilities to safeguard their devices, the more likely they are to comply with security standards. As such, the public sector should invest in information security training to create positive behavioral change among employees.

There are few studies examining the relationship between employee BYOD policy compliance and security culture. The study in this field is still in its infancy and is relatively new. D'Arcy and Greene [6] found some evidences of this association in a traditional computing context. This study aimed to determine if a security culture consisting of top management commitment and device monitoring could be a determinant of BYOD security policy compliance in the Malaysian public sector. However, results show that security culture did not have significant role on compliance. This could be attributed to the lack of understanding and involvement of senior management in organizational information security, as well as employees' interest in surveillance measures for personal devices. A security culture exists when employees know about security risks and how to avoid them, take responsibility, and take necessary steps to improve security. However, most importantly, the study provided empirical support for the relationship between SETA and security culture, which is mostly theoretical so far [14].

Systematic literature review found BYOD IT support, psychological ownership, and self-efficacy as important and under-researched factors to be included in the conceptual research framework. Although psychological ownership has been extensively studied in fields such as psychology, this factor has not been widely explored in the context of information security. This study offers a contribution to BYOD security policy compliance study by utilizing additional construct, the psychological ownership to gain insight into processes that affect employees' intention to comply security policies. As psychological ownership is shown to contribute significantly to the framework, the study provides practitioners and researchers with valuable insight into how to motivate employees effectively to comply with organizational security procedures. The sense of ownership makes users to be inclined to perform security behavior to keep information secure. Overall, the study reported that the R^2 value of the key dependent variable "security policy compliance intention" is 33.3%. This indicates that the framework is able to explain the factors influencing employees' BYOD security policy.

A significant contribution of this study is that it will provide a better understanding on the BYOD security policy compliance, particularly in the context of public sector in a developing country like Malaysia. To the best of knowledge, most of extant studies are based on individuals in western countries and there is still lack of research among Malaysian individuals. Implementing research that focuses on the local context is essential as the findings from the western context could not be generalized to other regions such as Asia, particularly in Malaysia, where it has a high rate of smartphone consumerization.

6.2.2 Practical Contributions

The results highlight several managerial implications.

a. **BYOD security risk awareness for the public sector**

This study produced a list of BYOD security risks emanating from security policy non-compliance from a systematic literature review. Performing risk assessment helps organization to gaining an understanding on threat severity, vulnerabilities and threat probability that can improve the effectiveness of information security in the CNII agencies particularly. It can create awareness among employees about the threat landscape of the organization to better comply with workplace security policy.

Organizations are not only able to identify security threats surrounding BYOD, but also the benefits to be taken into consideration when implementing BYOD. In order for businesses to maximize the benefits of bring-your-own-device policies, they need to have a comprehensive understanding of the associated risks [10].

b. **Prioritization of strategies based on security risks**

Evaluating risks encourages organizations to remain vigilant regarding the severity of each risk and, as a result, to coordinate strategies to meet these difficulties in accordance with their risk appetite. Evaluating risks prompts organizations to be vigilant on how critical each risk is and subsequently, coordinate strategies to address these specific challenges according to their risk appetite. Knowing where to begin to address security vulnerabilities enables organizations to maximize their IT resources and reduce expenditure. The findings provide important opportunity for CNII organizations to prevent issues in the workplace from escalating and becoming a potential threat to the availability of critical services. These issues are all the more important to be proactively considered in handling new contexts due to the Covid-19 pandemic, which made the workforce to work remotely. The pandemic situation has made employees to rely on their own devices and in most cases, more than one device to carry out their work duties.

c. **BYOD Security Policy Compliance Framework**

Successful ISP implementation is largely dependent on the government having the right tools at hand to identify compliance factors and to formulate a good policy based on them. The study identified and validated factors that influence employees' BYOD security policy compliance which serves as a guideline to policy makers. Additionally, given the high cost of remediating security breaches and incidents, this framework offers guidance to the management to improve security policy compliance (Fig. 6.1). These factors are crucial in promoting compliance behavior corresponding to the increased security policy requirements and the need to comply [5].

As the findings indicate that the act of specifying a security policy has a significant impact on mandatoriness and security policy compliance, management should formulate clear policies and convey them to employees to emphasize the significance of protecting information assets. It is important to develop organizational policies relevant to the new security challenges from BYOD. Technical and non-technical terms should be simplified for employees at all levels. They will be able to grasp their duties and responsibilities with clear instructions, reducing the likelihood of security incidents.

The significance of self-efficacy advises management on the need to invest in training programs for the employees. The benefit for this investment will be employees who have good understanding of BYOD information security will more willing to comply to security policies and at the same time improve the security culture at the work place. The security awareness initiative would be important for stimulating employees' attitude, while the training can offer specific BYOD security related knowledge and skills to improve security compliance in the organization.

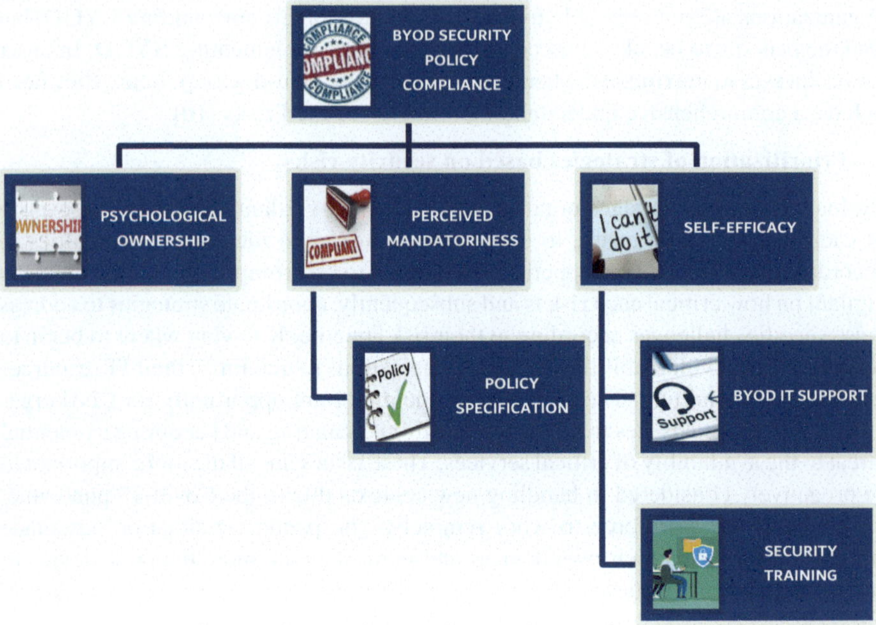

Fig. 6.1 Guidelines to improve BYOD security policy compliance

Awareness should be raised to ensure that employees do not forget their responsibilities to comply with ISPs' BYOD security policies and to be informed on information security threats.

Effectiveness of security training programs should be improved on a regular basis to satisfy the needs of self-efficacy. Management needs to design training programs that can effectively foster this efficacy belief. The threat evolves very fast that although it is difficult for employees to keep up with threats, a well-developed training program can narrow the gap. This research observes that existing studies on security culture is limited and this study found surprising results as security culture showed no significance towards compliance behavior. Interestingly, findings also highlight the importance of SETA as it affects security culture. Hence, organizations need to actively promote cyber security training and education in order to cultivate security culture.

The findings clearly elucidated the importance of having BYOD technical support to assist employees who use personal devices for work. Implementing a reliable technical support is vital to help employees use their devices securely while being productive.

6.3 Methodological Contributions

While quantitative and qualitative research are valuable in their own right, a mixed-methods approach is potentially superior because it can provide rich insights into the target phenomena and develop novel theoretical models [12]. This study employs qualitative and quantitative methods to analyze the factors influencing BYOD security policy compliance framework in order to increase the study's validity and reliability. In this study, qualitative study was used to explore and identify the factors BYOD security policy compliance. The qualitative approach is crucial in providing insights about the real issues on the ground besides offering in-depth analysis on BYOD security. Meanwhile, the quantitative study was deployed to test the relationship between the factors with security policy compliance. When conducting research, it might be helpful to combine multiple different methods in order to gain a deeper understanding of the issue being investigated. This approach proved beneficial in exploring the factors that influence BYOD security policy compliance from the population rather than solely being determined by extant research.

A review of literature showed that there is lack of studies exploring constructs from any other theories besides Protection Motivation Theory. Researcher was able to look at employees' compliance with the BYOD security policy from numerous angles after including these constructs in the framework for measuring policy compliance.

The quantitative analysis makes important contributions towards understanding employees' compliance intention towards BYOD security policy. The study has heeded the call to incorporate new methodological approaches to augment the understanding of employees' security motivation [13] and propose a theoretical framework by integrating factors that may influence BYOD security policy compliance [11].

Another significant methodology implication involves research methodology and the choice of statistical analysis in this study. This study is among the very few researches that utilized qualitative method (semi-structured interviews with information security experts and risks assessment to identify control measures that were translated into factors contributing to the framework). PLS-SEM was used to test the hypotheses and validate the framework that was developed based on the integration of theories to corroborate employees' intentions to comply with the BYOD security policy.

6.4 Study Limitations

This book has contributed novel findings, however there are some limitations that needs to be acknowledged. There are several caveats in this study that create opportunities for further research.

Data for the qualitative study were acquired from the perceptions of security specialists regarding their observations on BYOD practices. These views, however, may differ from employees' or ordinary insiders' perspectives on BYOD practices and security behavior. These two groups may have different abilities, knowledge, responsibilities, perspectives, and interpersonal interactions [1]. It would be effective to expand the study to include a larger sample of employees from various firms or industries. Including a variety of viewpoints will enrich this qualitative investigation.

First, the study is limited to selected CNII agencies in the Malaysian public sector. The approval procedure for conducting research in the Malaysian public sector was complex and time-consuming. It was even more challenging due to the pandemic, with majority of the workforce working remotely. Hence, due to the difficulties, this study could not cover all the other CNII agencies. Though the sample was appropriate for the PLS analysis, it still limits the possibility of generalizing the results and findings of this research.

This study demonstrated how factors affects employee's compliance intention for information security; however, this study did not directly measure employees' compliance performance or behavior regarding information security. Self-report measures of security behavior were used in the study. Social desirability bias can affect self-report measures, as people may describe what they think is acceptable rather than what they actually do. The current study has confirmed that CMV had no impact on the data, and that the data are valid.

Mediating or moderating effects were not tested in this study. Therefore, an important recommendation for future research is to apply moderating variables, such as gender, education level, job (IT/non-IT) and examine its impact on the framework.

6.5 Recommendations for Future Research

Based on the limitations mentioned above, this research provides several opportunities for future research, as explained below. Although it is plausible to assume that behavioral intention (compliance intention) can predict real conduct, future study may measure actual behavior to establish the relationship towards BYOD information security policy compliance.

Contradictory finding in this study was the insignificant relationship between security culture and BYOD security policy compliance intention. This could be due to the nature of security culture that is multilayered and multifaceted that requires immense effort and investment from all level of employees. To fully understand this non-significant relationship, additional investigations are required to gather on employees' perception of security culture in the public sector.

This study sheds some light on BYOD security policy compliance from the public sector CNII agencies' perspective. Though these agencies are supposed to be more focused on information security aspects, there is limited knowledge among employees on BYOD security risks and ISP compliance. This study may be extended to other agencies to confirm the findings.

6.6 Conclusion

Organizations face new and growing security challenges as BYOD continues to be integrated into organizational workflows. BYOD is a phenomenon that is here to stay. As the technology horizon continues to evolve making BYOD a working platform to reach customers and deliver service, the cybersecurity attack surface continues to increase, posing great challenges to organizations. Employees play a vital role in preserving the information security of their organizations; therefore, knowledge of BYOD security policy compliance factors can considerably enhance their conduct in this regard.

The research has achieved its objectives as stated in Chapter One and in Sect. 6.2 of this chapter. The objectives have been realized by conducting exploratory sequential mixed method approach. The study began with a systematic literature review to identify BYOD security policy compliance factors to establish the research framework. Following that, another systematic literature review was performed to gather and categorize risk according to People, Process and Technology model. Expert interviews were conducted to gather responses from security experts on BYOD practice and security risk. The experts also performed risk assessment to identify critical risk and recommend factors to mitigate the risk. The factors formed the conceptual framework. Next, survey was conducted among public sector employees to validate the framework. The findings were built upon the results gathered from qualitative and quantitative data analysis and confirmed that perceived mandatoriness, self-efficacy and psychological ownership did influence employees' BYOD security policy compliance, whilst security culture failed to show any support. This has also set the path for future research examining security culture-related constructs.

In conclusion, the main aim of the study has been attained and the findings help to extend the body of knowledge on the potential BYOD security risk and factors influencing BYOD security policy compliance. While contributing to the literature of human security behavior in information system, the findings will enhance the organizations' readiness in facing any security breach relating to BYOD. The findings are particularly beneficial to policy makers in improving organizations strategic plans in information security by emphasizing management and human issues in the public sector information security management.

References

1. Albrechtsen, E., Hovden, J.: The information security digital divide between information security managers and users. Comput. Secur. **28**(6), 476–490 (2009). https://doi.org/10.1016/j.cose.2009.01.003
2. Blythe, J.M., Coventry, L.: Costly but effective: comparing the factors that influence employee anti-malware behaviours. Comput. Hum. Behav. **87**, 87–97 (2018). https://doi.org/10.1016/j.chb.2018.05.023

3. Boss, S.R., Kirsch, L.J., Angermeier, I., Shingler, R.A., Boss, R.W.: If someone is watching, I'll do what I'm asked: mandatoriness, control, and information security. Eur. J. Inf. Syst. **18**(2), 151–164 (2009). https://doi.org/10.1057/ejis.2009.8
4. Clare, S.: 6 Advantages of BYOD and how it increases productivity and profits. Voxer (2019). https://www.voxer.com/blog/six-ways-byod-increases-enterprise-productivity-and-profits/
5. Cram, W.A., Proudfoot, J.G., D'Arcy, J.: When enough is enough: Investigating the antecedents and consequences of information security fatigue. Inf. Syst. J. **31**(4), 521–549 (2021). https://doi.org/10.1111/isj.12319
6. D'Arcy, J., Greene, G.: Security culture and the employment relationship as drivers of employees' security compliance. Inf. Manag. Comput. Secur. **22**(5), 474–489 (2014). https://doi.org/10.1108/IMCS-08-2013-0057
7. Greene, J.C., Caracelli, V.J., Graham, W.F.: Toward a conceptual framework for mixed-method evaluation designs. Educ. Eval. Policy Anal. **11**(3), 255–274 (1989). https://doi.org/10.3102/01623737011003255
8. Ng, B.Y., Kankanhalli, A., Xu, Y. (Calvin): Studying users' computer security behavior: a health belief perspective. Decision Support Syst. **46**(4), 815–825 (2009). https://doi.org/10.1016/j.dss.2008.11.010
9. Thompson, N., McGill, T.J., Wang, X.: "Security begins at home": determinants of home computer and mobile device security behavior. Comput. Secur. **70**, 376–391 (2017). https://doi.org/10.1016/j.cose.2017.07.003
10. Timms, K.: BYOD must be met with a wider appreciation of the cyber-security threat. Comput. Fraud Secur. **2017**(7), 5–8 (2017). https://doi.org/10.1016/S1361-3723(17)30058-1
11. Tu, Z., Adkins, J., Zhao, G.Y.: Complying with BYOD security policies: a moderation model based on Protection Motivation Theory. J. Midwest Assoc. Inf. Syst. **2019**(1), 11–28 (2019). https://doi.org/10.17705/3jmwa.000045
12. Venkatesh, V., Brown, S.A., Sullivan, Y.W.: Guidelines for conducting mixed-methods research: an extension and illustration. J. Assoc. Inf. Syst. **17**(7), 435–495. https://doi.org/10.17705/1jais.00433
13. Warkentin, M., Willison, R.: Behavioral and policy issues in information systems security: the insider threat. Eur. J. Inf. Syst. **18**(2), 101–105 (2009). https://doi.org/10.1057/ejis.2009.12
14. Wiley, A., McCormac, A., Calic, D.: More than the individual: Examining the relationship between culture and information security awareness. Comput. Secur. **88** (2020). https://doi.org/10.1016/j.cose.2019.101640

The manufacturer's authorised representative in the EU is Springer Nature Customer Service Centre GmbH, Europaplatz 3, 69115 Heidelberg, Germany. If you have any concerns regarding our products, please contact ProductSafety@springernature.com

Printed and bound by CPI Group (UK) Ltd, Croydon, CR0 4YY
30/09/2026
02225568-0002